Messerschmitt Bf109

Ground crew paints Oblt Hans-Jochim Marseille's 50th kill bar on his Bf109F. Gazala strip, desert, February 1942.

Classic WWII Aviation

MESSERSCHMITT Bf109

General Editor
Edward Shacklady

TEMPUS

First published 2000

PUBLISHED IN THE UNITED KINGDOM BY:

Tempus Publishing Ltd
The Mill, Brimscombe Port
Stroud, Gloucestershire GL5 2QG

PUBLISHED IN THE UNITED STATES OF AMERICA BY:

Tempus Publishing Inc.
2 Cumberland Street
Charleston, SC 29401

Tempus books are available in France, Germany and Belgium
from the following addresses:

Tempus Publishing Group	Tempus Publishing Group	Tempus Publishing Group
21 Avenue de la République	Gustav-Adolf-Straße 3	Place de L'Alma 4/5
37300 Joué-lès-Tours	99084 Erfurt	1200 Brussels
FRANCE	GERMANY	BELGIUM

British Library Cataloguing in Publication Data.
A catalogue record for this book is available from the British Library.

ISBN 0 7524 2003 8

Typesetting and origination by Tempus Publishing.
PRINTED AND BOUND IN GREAT BRITAIN.

Contents

Series Introduction

Many different types of aircraft were involved in the Second World War, new, old, conventional, and some, downright bizarre. Any selection in a series of 'classic' aircraft will, therefore, always be arbitrary and subject to the views of individuals. The selection of aircraft for this series has been primarily governed by their operational importance, although some types have, by necessity, other claims for inclusion. The series also seeks to cover a wide spectrum of the different operations involved during the conflict as well as those countries that had a leading role.

The major powers, during the 1930s, were becoming increasingly aware that the political instability throughout the world would inevitably lead to military conflict. Many of these countries had learnt lessons from the almost haphazard preparations of the First World War and were determined not to be in that position again. During this decade, particularly the middle and the late period, there was considerable activity for the manufacturers of aircraft as it was perceived that land battles would be of less significance as the fight, using heavily armed bombers, could be taken to the very doorstep of the enemy. Most countries looked on the development and production of aircraft as a defence measure, while others realised that a strong airforce would give them a total advantage over territorial claims – whether legitimate or otherwise. The advent of Adolf Hitler's rise to power in Germany, and the increasing strength of the new *Luftwaffe*, led many nations to the realisation that to rely on their air forces' existing capabilities would be extremely unwise and that they had to expand and re-equip with more modern combat aircraft. However, despite this obvious threat no country, at the outbreak of the Second World War, had the numerical strength or modern equipment to compare with that of the *Luftwaffe*.

The Spanish Civil War (1936-1939) afforded several of the major air powers, particularly Germany and Italy, an ideal opportunity to put their newly designed aircraft to the test under battle conditions. The pilots of Germany's *Legion Condor* and Italy's *Aviazione Legionaria* evolved a number of strategies that were utilised in the early part of the Second World War and the senior officers of the *Luftwaffe* were quick to realise the need for specialised ground-attack aircraft. On the other hand, the often inferior opposition and the ease with which they were eliminated, gave the German and Italian aircrew, as well as the officials of their respective air forces, an over-estimated view of the superiority of their aircraft.

The German Messerschmitt Bf109 had been conceived in the flush of Hitler's take-over of power in 1933, and as a monoplane had complete superiority in the air until the appearance of the Spitfire. Yet the Bf109E, the 'Emil', which provided the main fighter force for Germany during the first year of the Second World War, including the Battle of Britain period, had not evolved significantly from the MeBf109C which was the predominant fighter aircraft used by the *Luftwaffe* in the Spanish Civil War.

Italy's pilots had a totally different concept and still preferred the open cockpit and light armoury in the belief that it would enable them to out-manoeuvre their opponents.

During the immediate pre-war years the peacetime expansion of the Royal Air Force, by comparison with Germany, was slow and hampered by financial restraints. Like Italy, Britain was reluctant to dispense with their bi-plane fighters until the monoplane had proved itself. Although the manufacturers of Britain's two monoplane developments, the Hurricane and the Spitfire, were given substantial pre-war orders the RAF, by the outbreak of the Second World War, had little more than 300 Hurricanes in first-line service and approximately 150 Spitfires – less than a tenth of those ordered. In 1938, when war was very much in the offing, the RAF's weakness was only too apparent and a delegation, the British Purchasing Mission, went to the United States to order substantial quantities of US combat aircraft in an attempt to fill the gap. Most of these aircraft were not delivered until sometime in 1940 and, when war broke out, the RAF had to supplement their inferior numbers of Hurricanes and Spitfires with Gloster Gladiators and Fairey Battles which were no match for the 'Emil'.

The Soviet Union, who had also sent aircraft and pilots to the Spanish Civil War, was, in common with Britain, France and other European nations, still in the early stages of its modernisation programme. Although at the time of the German invasion the Soviet airforce was numerically strong its front-line aircraft were anything but modern. Like Britain, a few years previously, the Soviet Union had to rely on aircraft from the United States until they were able to produce, in sufficient quantities, their own.

Like the Italian pilots the aircrew of the Japanese airforce seemed to have a predilection for open cockpits and lightly armed but highly manoeuvrable aircraft. However, they discarded the bi-plane somewhat earlier and, at the time they opened hostilities against the United States, all principal first-line Army and Navy fighters were monoplanes, including the Mitsubishi Zero-Sen single-seater of the Japanese Naval Air Force.

At the time of Pearl Harbor, in December 1941, the United States' aviation industry was already producing large numbers of aircraft from orders placed on them by Britain, France and other countries. The already huge work load on the US aircraft industry was increased still further by the demand for production on behalf of its own forces. Nearly one-third of the entire US aircraft production was devoted to the manufacture of transport aircraft, with a high percentage of the remaining capacity involved with producing medium and heavy bombers. Nonetheless the overall output, from 1941 to 1945, included in excess of 12,000 Mustangs, 12,000 Corsairs, 15,000 Thunderbolts and 20,000 Hellcats and Wildcats, in addition to lesser quantities of other fighters.

The Classic Second World War Aircraft series is designed to give the aviation enthusiast a comprehensive history of many of the aircraft used during this period. Each title will cover the prototype development, production and operational use of the aircraft used by the airforces of Germany, the United States, Britain, France, Italy, Japan, the Soviet Union, and other countries that were involved in the conflict. The series will cover fighter aircraft and both heavy and medium bombers in narrative text, many black & white photographs and line drawings, and colour drawings that will show the different types of aircraft as well as the many colour schemes used by both squadrons and individuals.

INTRODUCTION

Messerschmitt's Immortal Bf109

When England and France went to war with Germany in 1939, for the second time in a generation, the horror of the first war was still vivid and remembered with dismay by those that had fought in the mud of Flanders. They imagined their sons would have to face the cruel misery of fighting in the ever present mud, and attempt to cross 'No-Man's' land in order to get to grips with the opposing German forces, who swept the advancing Allied soldiers away with the deadly machine guns.

But, those old soldiers were wrong as was soon proved when Hitler launched his *Blitzkreig* that subjugated Poland, the Low Countries and France. The German Panzer divisions, supported by hordes of dive and medium bombers, together with fighter aircraft, were more than a match for any of the nations that faced them.

England stood alone and awaited the cross Channel attack that Germany would have to initiate if it was to smash the last opposition that the Allied forces could muster. However, the more experienced of the German Chiefs of Staff knew that to cross that narrow stretch of water, that separated England from the Continent, the fleets of the Royal Navy would have to be destroyed. As Germany's navy did not possess a similar strength the only possible way to beat England was by ground forces and the *Luftwaffe*.

The chief of that arm was *Generalfeldmarschall* Göring, a bombast who informed his Führer that he could drive the Royal Air Force out of the air to allow the invasion of Britain under Operation *Sealion*. His Ju87 Stuka dive-bombers, Ju88s and He111s would be more than a match for the Royal Navy. With this enthusiasm and assurance Hitler delayed the invasion plans to allow Göring his opportunity and the latter launched his mighty *Luftwaffe* against England in the summer of 1940.

The Royal Air Force was not the force that went to war in 1939 as its fighters and medium bombers had suffered great losses during the Battle of France. Facing the *Luftwaffe* it could only call on two fighters, the Hurricane and the Spitfire, to defend England. The *Luftwaffe* had one similar fighter, the Messerschmitt Bf109, and while this aircraft could cope with the Hurricane, the Spitfire was more than its match with manoeuvrability and armament. However, there were more Bf109s than the combined numbers of the two British fighters and, until Göring ordered his fighter pilots to protect the bombers which were being badly mauled, there would have been little doubt but that the *Luftwaffe* would have had air superiority over Britain.

CHAPTER ONE

The Messerschmitt 109 Saga

To describe the history of the Messerschmitt Bf109 fighter as a saga could be said to be fanciful and invite criticism. But, in order to evaluate the fighter, it is necessary to review the early history of the company that ultimately developed it; its early successes and failures, and the role of Willy Messerschmitt himself.

The basic design had been established by a company that possessed no previous continuity of fighter evolution, neither had it any established resources of modern research. What was evident was that the company's design staff had embarked on an entirely new project with little previous experience in what was, at that time, an advanced technology.

The company had been formed on 30 July 1926 when the Bayerische Flugzeuwerke A.G. (BFW) was created by the *Reichsverkehrsministerium* (State Ministry of Transport), the Bavarian State Government and a merchant banking house. The new company took over the assets, and liabilities, of another defunct aircraft company that had produced, among other things, a series of designs that included the U11 Kondor transport which could accommodate eight passengers and a three-man crew.

Another successful type was the U12 Flamingo basic trainer and the new company purchased the factory at Augsburg to develop the U12 design. This eventually attracted orders from Hungary, which was supplied with twenty-four completed aircraft and an agreement for the licence production of the aeroplane. Of this total Austria purchased ten and it was also built under licence in Latvia.

Despite these early successes the new company lacked the expertise of a competent designer and the owners, therefore, cast around for a suitable candidate. He appeared in the form of Willy Messerschmitt who was, at that time, employed by his own company, Messerschmitt Flugzeugbau GmbH. This company was building the successful light transport monoplane, the M18, in addition to building many sailplane designs.

Messerschmitt was persuaded to merge his company with BFW, and part of the agreement between the two was that Messerschmitt would supply the company with all its design requirements which would allow BFW to concentrate on developing and building these designs. A contract was placed with BFW for the production of twelve M18s, in an all-metal version, for a new domestic airline and Messerschmitt successfully negotiated a subsidy from the Bavarian State Government.

The next Messerschmitt design was the M19 two-seat sports monoplane and it was the first new Messerschmitt design to be produced by BFW. He also designed the M20 transport aircraft which crashed during its first flight. As a result Deutsche Lufthansa (DLH), a transport company, cancelled an order for the aircraft, but, after a second prototype had successfully passed through it first flight and subsequent trials, the order was reinstated.

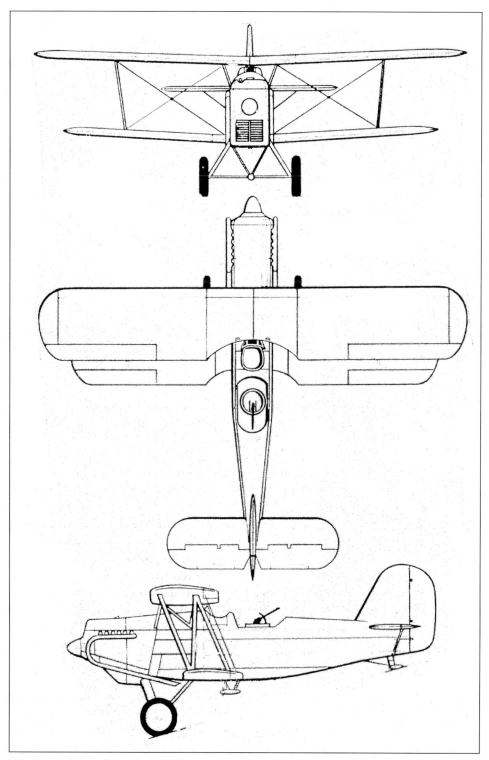

The Heinkel He45C was built by Messerschmitt under licence from Heinkel.

The company was, in the meantime, sold to an entrepreneur with Messerschmitt retaining a share holding. He was appointed as the manager with the entrepreneur as chairman. BFW was now building a whole range of aircraft including the M20b, M21 trainer, the M22 bomber, a twin-engined craft which had been ordered by the Ministry of Aviation, the M23 sports monoplane and the M24 eight seat transport. The M22 was cancelled after the prototype crashed but the M23 was a successful design. However, BFW was overstretched with so many irons in the fire that it was financially unsound to the extent that Deutsche Lufthansa cancelled the M20b contract, which was almost complete with all aircraft at an advanced state of construction.

As a result the company was wound up in 1931 but there was hope for Messerschmitt. He had retained all patents and designs and had also managed to raise additional capital. Also DLH was forced to accept delivery of the M20b aircraft in addition to a second contract for the M29 mail-plane. Things were a little brighter for Messerschmitt as he had also negotiated a contract for the design of a new, two-seat, sports aircraft.

In May 1933 the new company, also named Bayerische Flugzeugwerke, was formed with Messerschmitt acting as co-manager. The only drawback to all this activity was when Erhard Milch, the former managing director of Deutsche Lufthansa, was appointed State Secretary of Aviation, for it was he that had cancelled the contract for the M20 transport aircraft. He had come to dislike Messerschmitt, during the course of their relationship, and revealed it by attempting to restrict the new company's role to that of a licensed manufacturer of other companies' designs.

However, this did enable Messerschmitt to consolidate the company on a firm basis as they had contracts for thirty Dornier Do11s and twenty-four Heinkel He45s.

At this stage of development the company acquired a new designer, Walter Rethel, who had been an Arado employee and had designed the M35 sports monoplane and M36 eight-seat transport for Romania.

The company then received a contract to design a competition aircraft for participation in the *Challenge de Tourisme Internationale* of 1934. When Milch learned of Messerschmitt accepting this contract he was furious and his relationship with Messerschmitt worsened. However, the M37 was adapted to meet this new requirement and the prototype was named the Bf108 Taifun, or Typhoon, of which six examples were ordered. The prototype flew in the spring of 1934 and was an all-metal, light aircraft of advanced engineering.

It had a flush riveted stressed-skin airframe and a retractable undercarriage. A small wing area, embodying the Messerschmitt patented single-spar construction with Handley-Page leading edge slots and training edge flaps with which to obtain the necessary lift. Flight trials proved that the design was exemplary.

Having proved to be capable of designing and producing such advanced aircraft the company was to receive contracts to build yet more Dornier Do11 twin-engined bombers and Heinkel He45 biplane fighters, but no original work was approved.

The *Luftwaffe* was expanding rapidly and Milch of the *Reichsluftfahrtministerium* (RLM) could no longer ignore the company as Germany was re-arming and required all the modern aircraft it needed to equip its squadrons. Milch did his utmost to deny Messerschmitt any entry to the field of designing, developing and building a military aircraft of his own design but once again he was over ruled and was forced to allow

Messerschmitt the opportunity to submit a design for a fast, high speed, military fighter.

Other well known and established companies that tendered to this specification were Arado with their Ar80, Focke Wulf and their parasol Fw159 and Heinkel and the He112. All received contracts for a single prototype.

Messerschmitt's history gave no hint of his capabilities in this field, but he had anticipated the contract by initiating the design of his first fighter, the Bf109 before the first flight of the Taifun.

He chose to design the airframe as small as was possible to provide room for the pilot, a wing of sufficient area to lift and carry a suitable armament and an airframe that was strong enough to cope with the stress demanded of a front line fighter. Performance was the *raison d'être* for a fighter and Messerschmitt chose to retain the small area wing with its Handley-Page leading edge slots and flaps. The basic design was established despite the company having little, or no, previous experience of fighter aircraft development. Even more important it had no established resources of modern research. Messerschmitt was gambling that the company's design staff would accept the challenge of working on an entirely new project with little previous experience in advanced technology.

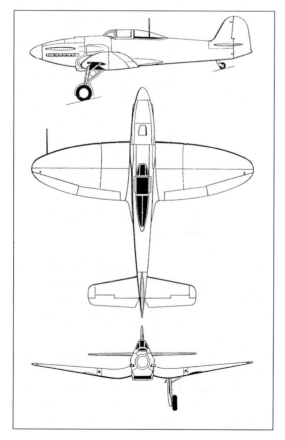

Another licence build by Messerschmitt was the Heinkel He112-B.1. This was submitted to the same specification of the Bf109 and although not as efficient as the latter was granted a contract. Small batches were built and extensively used by the Luftwaffe *as a propaganda weapon in addition to fighting in Spain on the Nationalist side.*

Prototype Bf109 V1, 109A as rolled out in August 1935.

The manufacture of the first prototype, the Bf109 VI, had retained a number of features of the 108 such as the slotted landing flaps, leading-edge (Handley-Page) flaps, retractable undercarriage and a single-seat adaptation of the enclosed cockpit. The fuselage was also an oval section, stressed-skin, metal monocoque with flush riveting, and the single spar wing was of similar structure. Work commenced in 1935, and aircraft (Werk Nr 758, registered D-AIBI) was completed in September with a 695hp Rolls-Royce Kestrel V engine owing to the intended 610hp Junkers Jumo 210A not being available.

Preliminary flight trials were undertaken by test pilot Knoetsch before the VI was delivered to Rechlin, where the undercarriage collapsed on landing. Repairs were made and the VI was delivered finally to Travemünde in October for competitive evaluation along with the He112 VI, the Ar80 VI and the Fw159 VI. That the Bf109 was declared the joint winner (with the He112) aroused astonishment in aviation circles – not shared by BFW, who had already started work on two further prototypes (V2, D-IUDE, Werk Nr 809, and V3, D-AHNY, Werk Nr 810).

The accident involving the undercarriage collapse was not recognised as being significant, but there was one design weakness in the Bf109 and this lay in the main-wheel gear. Had the designers incorporated a fundamental redesign of the landing gear in the second and third prototypes, and in production aeroplanes thereafter, the 109 would not have been continually dogged by undercarriage collapse in the hands of less experienced pilots.

In January 1936 the Bf109 V2 flew for the first time, to be followed by the V3 in June of the same year. By comparison the only Hurricane prototype flew in November 1935 and the Spitfire prototype in March 1936. All Service and manufacturer's trials were performed by these two prototypes and production examples did not appear until late summer of 1937 and 1938 respectively.

The 109V2 and V3 were both powered by Jumo 210A engines and had provision for two nose decking-mounted MG 17 rifle calibre machine guns that were synchronised to fire through the propeller disc. In retrospect this apparently light armament was understandable as being the traditional fighter armament that had persisted in the majority of fighters since the First World War. Rumour had it that the Hurricane and

Prototype Bf109 V4, D-IOQY.

Spitfire would eventually appear with four machine guns, and, at the last moment, it was decided to produce the Bf109B with three guns, the third firing through the propeller hub. This was to be replaced, at a later date, by a MG FF/M 20mm cannon when production supplies materialised. In the event the twin gun Bf109A never reached production status.

When the true armament of the Hurricane and Spitfire was finally disclosed as eight machine guns it was probably too late to include any substantial increase in armament in the Bf109. Nevertheless, the range and penetrating power of the single 20mm cannon probable tended to even up the 109's armament more than was accepted at the time.

The Travemünde trials resulted in an award of a contract for ten pre-production Bf109s, with BFW continuing to produce further prototypes. The Bf109 V4 (D-IOQY, Werk Nr 878) was the first of the prototypes armed with three machine guns, and flew towards the end of 1936. The V5 and V6 flew early in 1937, and were followed by the V7 – this latter prototype being representative of the production version, the Bf109B.

Bf109 V4 with strengthened wings.

Prototype Bf109 V7, D-IALY with VDM-Hamilton VP metal propeller.

CHAPTER TWO

Bf109B, C and D Variants

The pre-production batch of the Bf109B-0 was built during the spring of 1937, and these aircraft were delivered to various test establishments in Germany where they were flight tested by pilots who only had experience in flying the biplanes of early *Luftwaffe* years. The production Bf109B-1, powered by the 635hp Jumo 210D driving a fixed pitch, 2-bladed propeller, was restricted to a maximum speed of 292mph at 13,100 feet. German engine manufacturers were unable to produce an in-line engine at that time that compared to the Rolls-Royce Merlin, which was now producing 1,000hp and enabling the Hurricane Mk I to reach maximum speeds of approximately 325mph, with the Spitfire ahead of that.

Production B-Is entered service initially with the *Richthofen Jagdgeschwader* but the relatively poor performance, mainly attributed to the fixed-pitch propeller, gave rise to frustration among Service pilots who recognised, despite this, the potential of this beautifully handling fighter. The manufacturer had already acknowledged this shortcoming and had entered into a licence arrangement to fit sub-contracted Hamilton variable-pitch 2-blade propellers to the 670hp two-stage supercharged Jumo 210G, thereby producing the Bf109B-2 variant. At a loaded weight of 5,180lb this version possessed a maximum speed of 302mph at about 17,500 feet.

As previously mentioned the first *Luftwaffe* fighter squadron to receive supplies of the Bf109B-1 was *Jagdgeschwader 132 Richthofen* with plans to equip *II Gruppe* and then *I Gruppe*. As part of the strategy to gain combat experience it was decided that *Jagdgruppe 88*, then serving in Spain with the Condor Legion and flying He51 biplane fighters, would also be equipped with the Bf109B-1. The new fighters arrived in Spain in April 1937, and were issued to *2 Staffel/Jagdgruppe 88*. Relatively poor performance, mainly attributed to the fixed-pitch propeller, frustrated *Luftwaffe* personnel but this was overcome when the Hamilton variable-pitch 2-blade propeller was fitted.

If the Messerschmitt fighter was to be on a par with the RAF's new Hurricane and Spitfire a new engine was urgently required and one was under test in the 109V13 prototype. This was a standard B-series airframe fitted with a ground-boosted 1,650hp Daimler-Benz 601 engine and, in order to impress world aviation, it was flown by Dr Hermann Wurster on 11 November 1937 who established a new world land-plane speed record of 379.39mph. The Royal Air Force unofficially overtook this record when Squadron Leader John Gillan flew from Edinburgh to London in a standard Hurricane of No. 111 Squadron, RAF, at an average ground speed of over 400mph with a Rolls-Royce Merlin of 1,050hp.

Not to be beaten Germany made a new attempt for the Air Speed Record on 26 April 1939. A time of 469.22mph was recorded with a specially tuned Me109R. The British Air

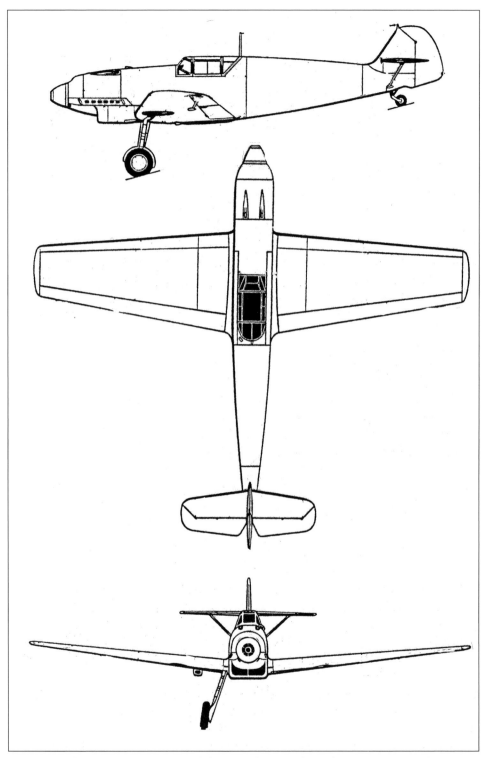

First series Bf109 was the 109B-1 which fought in the Spanish Civil War with the Condor Legion.

Bf109B-2 with Jumo 210G fuel-injected engine.

Bf109B-2, summer 1937.

Prototype Bf109 V13 with 3-blade metal VDM Hamilton propeller.

Ministry's response was to issue Specification F.35/35 for a 'Special' High Speed Racer/fighter which attracted entries from Airspeed, Bristol Aircraft Company, General Aircraft and Supermarine, the latter preparing a modified Spitfire powered by a highly tuned Rolls-Royce 'R' engine. This was to become part of the development programme for the Griffon engine. However, the Spitfire never competed and was refurbished as photo-reconnaissance variant.

The Condor Legion Bf109s.

By May 1937 about fifty Bf109B-1s and B-2s had entered service with the *Luftwaffe* (some of the early B-2s being provisionally fitted with Jumo 210E engines), and it was at this time that events abroad were to provide the opportunity to test new fighter under combat conditions.

The Spanish Civil War had been raging for a year, during which time the *Legion Condor* had been fighting with the Heinkel He51 and the Arado Ar68 biplanes. Russia supported the Republican Government of the day and was supplying their air force with the stubby 1-15 and 1-16 fighters which had obvious superiority over the biplanes. Not wishing his *Condor* pilots to be flying inferior aircraft to the communists, Göring, with Hitler's permission, equipped the 1st and 2nd *Staffeln* of *Jagdgruppe 88* with twenty-four Bf109B-1s and B-2s. The He51s were withdrawn from a pure fighter role and were to undertake ground attack duties with the 109s, at times, providing a support role.

Bf109B-1 of 2./JG88 Condor Legion, Spain 1937.

Many pilots that fought the Spanish campaign went on to chalk up impressive kill rates in the Second World War and became well known. The pilots included:

Pilot	Victories in Spain	Victories in WWII
Werner Mölders	14	101
Wolfgang Schellmann	12	14
Reinhard Seiler	9	102
Walter Oesau	8	115
Otto Bertram	8	13
Hans-Karl Mayer	8	24
Herbert Ihlefeld	7	123
Wilhelm Balthasar	7	33
Horst Tietzen	7	20
Walter Grabmann	6	6
Herbert Schob	6	22
Günther Lützow	5	103
Joachim Schlichting	5	3
Hubertus von Bonin	4	73
Wolfgang Lippert	4	25
Hannes Trautloft	4	25
Josef Fözö	3	24
Heinz Bretnütz	2	35
Karl-Wolfgang Redlich	2	41
Theodor Rossiwall	2	15
Wolfgang Ewald	1	77
Rudolf Resch	1	92

By and large the Messerschmitt 109s acquitted themselves well in Spain, proving themselves indisputably superior to all other combat aircraft in the theatre. An isolated instance of tail failure, which caused a fatal accident, was magnified by propaganda to suggest that the aircraft would break up in high altitude manoeuvres. Some weaknesses were disclosed, such as wing flutter (probably resulting from inadequately balanced ailerons) and airflow instability over the tail unit.

The light armament still drew criticism from the *Legion* pilots and one or two of their Bf109B-2s were fitted with the hub-firing cannon, but even this proved unsatisfactory owing to the guns jamming from wide-tolerance ammunition – this also occurred, at critical moments, with the Spitfire. At Augsburg BFW had prepared a new prototype, the V8, which featured two additional MG 17 guns in the wings, but these only added to the wing flutter troubles until the ailerons were redesigned.

In August 1938 Germany's first ever aircraft carrier, the *Graf Zeppelin*, was ready for launch and the offensive equipment specified was a navalised version of the Ju87 Stuka dive-bomber. *4 Trägergruppe/186* was equipped with these with a second unit, *6*

Bf109B-2 of 2./J88, Spain 1938.

Tragergruppe/186, to be equipped with a naval version of the Bf109B, as was *5 Tragergruppe/186*. The carrier was launched on 8 December 1938 but was doomed never to see service. Without a protective screen of other warships it would have proved to be just as vulnerable as HMS *Glorious* during the Narvik campaign.

A new prototype had also appeared, the V8, with two additional MG 17 m/gns in the wing. This was followed at the end of 1937 by the prototype V9 with an armament of two MG FF 20mm cannons in the wings in addition to the two nose-mounted MG 17s. It was being viewed as the prototype Bf109C which now could equal, or surpass, the same weight of ammunition as the eight-gun Hurricane and Spitfire.

A small pre-production batch of Bf109C-0s and the first production C-1s were armed with two MG 17s in the wings and two in the nose decking, while the C-2 had an additional hub firing MG 17. The C-4 also featured a hub-firing MG FF cannon and twelve Bf109C-1s were delivered to *3 J/88,* commanded by *Hauptmann* Werner Mölders.

Bf109B-2

Leading particulars

Wing span: 32ft 4.5in, area 176.53sqft. *Length:* 28ft 1in *Height:* (tail down over prop) 11ft 2in. *Weights:* tare 3,318lb, gross 4,740lb. *Max speed:* 255mph @ S/L, 260mph @ 3,280ft, 289mph @ 13,120ft, cruise 217mph @ 8,200ft. *Ceiling:* 26,900ft. *Height to time:* 19,685ft in 9.8mins. *Range:* 405 miles. *Engine:* Junkers Jumo 210Da of 700hp @ T/O, 675hp @ 12,500ft. *Armament:* 3/4 7.9 MG 17 m/gns, 700rpm (in wings and fuselage).

Bf109B-2 2./J88 Spain 1938.

Bf109C at training school.

The Bf109D Series

A prototype, V10 with a standard B-2 airframe, had been fitted with a Daimler-Benz DB200 Series engine of 960hp in August 1937, and was followed by the prototypes V11 and V12 with production DB600A engines. Although the type was referred to as the Bf109D it was always regarded as an interim variant until production of the Bf109E Series with the more powerful DB601 engine.

A limited pre-production batch of Bf109D-0 fighters was ordered by the RLM in August 1937, and series production of the 109D commenced in November and appeared in early 1938. Deliveries of the D-1 started soon after to *I Gruppe des Jagdgeschwaders 131.* The D-1 was similar to the previous C Model apart from local strengthening, in particular the main undercarriage and in having a 3-blade metal CP propeller. Armament was two 7.9 MG 17m/gns with 500rpg and a MG FF/M cannon (160 rounds). A number of aircraft dispensed with the hub mounted cannon and had increased volume of 1,000rpg for the m/gns.

The Bf109D-2 followed with wing mounted MG 17s making a total of four, and the D-3 with a pair of wing mounted MG FF 20mm cannon. A total of 200 Bf109Ds were delivered and it was this variant that fought the Polish campaign in addition to being used as night fighters over Germany. They were phased out in June/July 1940. Ten were sold to Switzerland and three to Hungary.

Bf109D-1

Leading particulars

Span: 32ft 4in, area 176.53sqft. *Length:* 28ft 2in *Height:* 8ft 4.5in. *Weights:* tare 3,964lb, gross 5,335lb. *Max speed:* 298mph @ S/L, 357mph @ 11,500ft. *Ceiling:* 32,100ft. *Rate of climb:* 2,985ft/min (initial). *Range:* 348 miles @ 11,600ft. *Engine:* Daimler Benz 986hp @ T/O, 910hp @ 13,000ft. *Armament:* one 20mm MG FF/M cannon with 160 rounds and two 7.9mm MG 17 m/gns with 500rpg.

Encouraged by the DB600's early promise, the Daimler-Benz engineers produced the superb 601 Series with increased supercharging and direct fuel injection. The latter made possible the dispensing of the G-prone carburettor, and permitted negative-G aircraft manoeuvring without engine cutting.

A small batch of 109D-0s were ordered in 1937 and the majority used for test aircraft of the forthcoming E Series such as Bf109 V14 and V15. Production Bf109D-1 was initiated and deliveries to *I Gruppe des Jagdgeschwaders 131* began in early 1938. They had minor modifications and a VDM 3-blade CP metal propeller. Fuel capacity was increased to 88gals. Armament was now two 7.9mm MG 17 m/gns, 500rpg and one 20mm MG/FM cannon with 160 rounds, with all concentrated in the nose section.

The Bf109D-2 had two additional MG 17s in the wings, while the 109D-3 had only two wing-mounted MG/FF cannon. The variant fought in the Polish campaign and France.

Prototype Bf109 V14 D-IRT. Pre-production model for the D Series.

Bf109D-1

Leading particulars

Wing span: 32ft 4.5in, area 176.53sqft. *Length:* 28ft 2.5in. *Height:* 8ft 4.5in. *Weights:* tare 3,974lb, gross 4,398lb, max permissible 5335lb. *Max speed:* 298mph @ S/L, 357mph @ 11,480ft. *Ceiling:* 32,810ft. *Rate of climb:* 2,985ft/min @ S/L *Range:* 348 miles @ 11,810ft. *Engine:* Daimler Benz DB 600Aa of 986hp @ T/O, 910hp @ 13,210ft. *Fuel:* 88gals. *Armament:* one 20mm MG FF/M cannon 160rpg and two 7.9mm MG 17 m/gns with 500rpg.

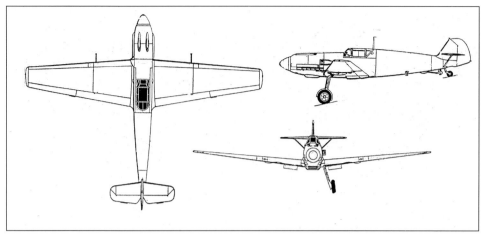

The Bf109T-1 carrier borne fighter intended for the German Navy aircraft carrier Graf Zeppelin. It was never used as the carrier was never completed, and all production aircraft were converted to the T-2 standards as fighter bombers.

Bf109B-1.

Bf109C-2, I./JG 137.

Bf109C-2 at training unit, 1940.

Bf109C-2, II./JG 71 being re-armed and refuelled at Friedrichshafen, 1930.

CHAPTER THREE

The Ubiquitous 'Emil'

The 109E was virtually a direct development of the 109D V14 (D-IRTT) and V15 (D-IPHR) pre-production prototypes, and the first named first flew in mid-1938 powered by a 1,100hp Daimler-Benz DB601A with direct fuel injection and improved supercharging of 1,050hp @ T/O and 1,000 (2,400rpm) @ 12,200ft. Armament consisted of two MG 17, 7.9mm nose machine guns and two MF/FF 20mm wing cannons. The V15 which followed was the test-bed for the single, hub-firing cannon which replaced the wing guns. A small pre-production batch of ten E-0s was completed in December 1938. This was followed by full-scale production the following year.

The Emil, as the Bf109E was popularly known throughout the *Luftwaffe*, was the first variant to reach major production status with the first production models coming off the lines in early 1939. Such was the demand for the new version that a new factory had to be built solely for its production in addition to the current Regensburg factory. Production rates soared to the position whereby a total of 1,540 were produced in 1939.

First deliveries were made in February of that year with fifteen early E-1s being despatched to Spain in April 1939 – the serial numbers ranged between 6.100 and 6.130. They were armed with four MG 17s as the engine hub mounted cannon was still considered to be unreliable, or two MG/FF 20mm cannon, with 60rpg in the wings.

The E-1/B was a sub-variant equipped with light bomb racks holding four SC50 (110lb) bombs or one of 550lb. Due to Messerschmitt's production of the new Bf110

Cannon armed Bf109E-1, 25 Grupo, 23 Regimiento de Saza, Spanish Air Force.

Bf109E-1s were operated by the Swiss Air Force.

Zerstörer fighter, production of the 109 was transferred from Augsburg to Regensburg, two factories that were to be long remembered among Eighth Air Force bomber crews.

The German fighter still had had faults with light and sensitive controls in the lower speed range that tended to stiffen above 300mph, leading to pilot fatigue in combat. The Hurricane and Spitfire could out-turn the Messerschmitt with the Hurricane in its element at the altitude band between 12,000 and 17,000 feet. When employed in the close escort role for the German bombers, which tended to fly in this band, any performance advantage that the Emil possessed over the Hurricane was rendered relatively unimportant.

The test bed for the Bf109E-3 was the V17 with an improved DB601A engine which had provision for a single 20mm MG FF/M cannon on the crankcase and firing through the propeller hub. This engine provided 1,175hp @ 14,765ft and 1,020hp @ 14,700ft. The E-3 replaced the E-1 in late 1939 but took little, or no, part in the invasions of Denmark and Norway. Unfortunately for the *Luftwaffe* one was captured intact in May 1940 at the beginning of the Battle of France and tested at Farnborough as AE479.

Into Battle

Just as Britain feverishly worked to modernise her metropolitan fighter squadrons by accelerated introduction of the Hurricane and Spitfire, the *Luftwaffe* strove to introduce the Emil as its standard interceptor and by September 1939 twelve *Gruppen* had been equipped up to a Unit Establishment of 850 aircraft. In view of the likely character of the forthcoming Polish campaign, and displaying the tactical policies of *Blitzkreig*, these *Gruppen* were to be employed in what was, twenty years later, to be called the 'Air Superiority' role, in effect ground support and battlefield cover. Approximately one in eight of the Emils deployed carried bombs and were designated E/1/Bs. Although they were administratively organised into *Jagdgeschwaderen* (i.e. JG1, 2, 3, 26, 52 and 53), they were autonomous ground support units and employed tactically as such. A high number, 235, of obsolescent Bf109Ds were also employed in the *Zerstörer-geschwaderen*.

Despite a gallant resistance by the Polish Air Force the campaign was quickly over, and the nature of the air fighting provided little in the way of constructive experience for the *Luftwaffe*, other than to reveal a number of logistic weaknesses. In the twenty-six day period of the Polish campaign the *Luftwaffe* suffered the loss of 203 aeroplanes destroyed, of which fewer than twenty were Bf109s.

Staffelkapitän, 4./JG 53, and pilots at Mannheim, Sandhofen Airfield. November 1939, with a Bf109E-3.

Bf109E-1, II/JG54 'Grünherz', 1939.

By the end of 1939 no fewer than 1,540 Bf109s had been completed, the majority at sub-contracting factories. In July 1938 the Bayerische Flugzeugwerke had been re-named Messerschmitt AG (although the 109 still carried the prefix 'Bf' throughout the war), and only about 150 aircraft had been completed at the Augsburg plant. Pressure of work on the larger Bf110 (*Zerstörer*) resulted in production of the Emil being switched to Regensburg-Obertraubling and Regensburg-Prufening. The Ago plant at Oschersleben, Arado at Warnemunde, Erla at Leipzig, Fieseler at Kassel and the WNF plants at Weiner-Neustadt and Delitzsch were engaged in sub-contracting components in a dispersal programme similar to the British 'shadow' factory plan.

Two new variants of the Emil were delivered in the months following the capitulation of Poland, the E-2 and E-3. Relatively few of the former reached operational units, but manufacturing records do suggest that the E-3 was built in greater numbers than any other version of the E-Series. This supposition is by no means borne out by existing *Luftwaffe* records, and almost certainly confirms that this version's provision to mount the hub-firing MG FF cannon, in addition to the two nose and two wing MG 17s, was never implemented, and the aircraft taken on charge by the Service were probably E-1s. By way of confirmation the tally of Bf109s published in Vol II of *Obd.L. Genst.Gen.Qu./6 Abt. Nr 2990*, and issued in March 1940, discloses that between them the *Jagdeschwaderen*, *Zerstoreschwaderen* and *Lehrgeschwader 1* had the following number of aircraft available:

Operational: 62 Bf109D, 444 Bf109E-1, 71 Bf109E-3.

Non-operational: 54 Bf109D, 61 Bf109E-1, 36 Bf109E-3.

Prototype of the 109E-3 can be considered to be the 109 V17 (D-IWKU) with the DB

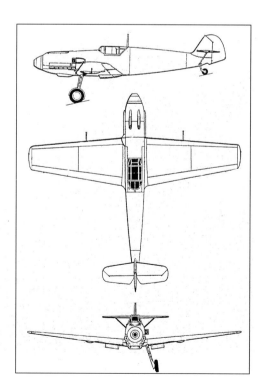

The Bf109E-3 Emil was the principal Luftwaffe fighter deployed during the Battle of Britain.

601Aa engine and its hub mounted 20mm MG FF/M cannon. It was during the Battle of France that consideration was given to using the E variant as a fighter bomber (*Jagdbomber*, or *Jabo*) and trials were conducted with a number of bombs weighing from 110 to 551lb on a ventral, fuselage rack. Operations were launched against British Channel shipping in the month preceding the Battle of Britain.

A number of Es were retrofitted with racks and bombs and carried the designation of 109E-1/B and normally carried a single 110lb bomb. The next retrofit 109 was the E-4/B with racks carrying either four 110 or one 551lb bombs. To acquire current information of Fighter Command's forces the 109E-5 was produced as a reconnaissance fighter. This normally had the wing-fitted MG FF cannon removed and a vertical Rb 50/30 camera installed on the rear fuselage.

The Bf109E-4/N that followed had the new version of the DB engine, the 601N installed. With 96 octane fuel and increased compression ratio it produced 1,200hp for take-off. The 109E-6 was a tactical reconnaissance development to replace the E-5.

It appears that the E-3s tended to be concentrated in the *Jagdgeschwaderen* responsible for home defence duties, and it was the aircraft of III *Gruppe*, JG77, that intercepted the

Bf109E-4, with and without ventral rack and bomb.

RAF Wellingtons of Nos 6, 39 and 147 Squadrons on 18 December 1939 when making their disastrous raid on Wilhemshafen. Twelve bombers were shot down for the loss of two Emils. One of the German pilots, *Leutnant* Johannes Steinhoff, shot down two Wellingtons and later went on to become one of the war's high scoring pilots with a final tally of 176 victories.

The Battle of France

With the end of the so-called 'Phoney War' six *Fliegerkorps* consisting of sixteen *Gruppen* equipped with Bf109E-1s and E-3s, distributed amongst eight *Jagdgeschwaderen* – JG2, 3, 26, 27, 51, 52, 53 and 54, with a total of 1,016 Bf109 fighters in May 1940 for the assault on the Allies in France. When the air operations against the Western Allies opened on 10 May the 109E swept all Allied aircraft before them and accounted for approximately sixty per cent of all British, Dutch, Belgian and French aircraft destroyed in the air. In the first day's fighting Bf109 losses amounted to six with the German fighters destroying at least eleven Dutch, eighteen Belgian, twelve British and around twenty French aircraft.

It was immediately recognised that the Hawker Hurricane was the only Allied fighter that came near to matching the Emil's performance, yet such were the desperate straits of the ground forces that few Hurricanes could be released for freelance interception duties. Their main role was established as providing escort for British light bombers attacking bridges and other key points being overwhelmed by the advancing German armies. Such was the strength of the *Luftwaffe* that the Hurricane escorts were quickly scattered, and by the evening of the 15 May no fewer than seventy-three RAF Hurricanes, in addition to numerous others of the Belgian Air Force, had been destroyed.

The French Air Force fared much worse, equipped as it was with a motley collection of obsolescent aircraft such as the Bloch 152s and Morane-Saulnier 406s, which were no match for the German Emils and suffered accordingly. The Dewoitine D520 came near the quality of the Bf109E-1, but only two *Groupes de Chasse* were being equipped with the type when the blow fell on 10 May. The *Armee de la Air* had recently taken delivery of the Curtiss radial engined Hawk (P-36), and the American company claimed that it could take on the German fighter on terms. Most, for some inexplicable reason, were transferred to the French colonies in Africa.

Although *Luftwaffe* records of the fighting between 10 May and 25 June are not quite complete, it is clear that a total of 583 Bf109Es were flown on operations, of which about seventy were lost or written off; and a further fifty-odd were badly damaged. As far as can be ascertained from surviving Allied records, some five or six hundred aircraft fell to the guns of the *Jagdgeschwaderen*. At least three Emils force-landed behind Allied lines at this time, either wholly intact or with only superficial damage. One, a Bf109E-3, came down near Amiens on 2 May and was flown briefly by a French pilot before being taken over by the RAF. After a number of useful flights by pilots of Nos 1 and 73 Squadrons in France, it was flown back to Boscombe Down and, subsequently, Farnborough where it underwent thorough evaluation.

An A&AAE Report, issued in July 1940, confirmed suspicions already voiced by British Hurricane pilots that the Emil possessed an important combat weakness – that of flicking

Bf109E-3, Jagdgeschwader 53 *'Pik As' (Ace of Spades), 1939.*

Bf109E-1s of II./JG 54. Early splinter camouflage 1938-39.

out of a tight turn after the leading edge slats opened. Obviously, experienced *Luftwaffe* pilots were well aware of this phenomenon and were careful not to be taken unawares, but the information was undoubtedly broadcast among British fighter squadrons whose Hurricanes possessed superb turning capabilities. This fact probably enabled the RAF to meet the Emil on more equal terms during the Battle of Britain.

Lest the foregoing gives the impression that the German fighters had matters all their own way throughout the Battle of France, it is necessary to outline the events of Operation *Dynamo*, the nine days that witnessed the evacuation of the British Expeditionary Forces from the beaches of Dunkirk. It was in the skies over Dunkirk that home-based Spitfires of RAF Fighter Command encountered the Bf109E. Almost at once the fortunes of air combat, if not reversed, were at least evened. The German fighters were now employed in close escort of bombers and were thus at a disadvantage, especially when accompanying the Ju87 dive-bombers. Great damage was done by the German bombers while their escort strove to keep the British fighter squadrons from interfering. Losses of aircraft during *Dynamo* have recently been stated as 106 British fighters and 71 other aircraft, compared with 132 German aircraft of all types. Nevertheless the close margin of losses was of little consequences compared with the survival of 338,500 British and French soldiers who returned to Britain.

Bf109E-4/B, 10 (Jabo) JG 27, North France, 1941. SC250, 550lb bomb.

Bf109E-4/B with two MG 17 wing and one hub cannon.

Bf109E-1, IV./JG 132, late 1939. Note German cavalry in the background.

Bf109E-1, I./JG 51. Note white hand motif on fuselage.

CHAPTER FOUR

The Battle of Britain

The triumphant German campaign in France was no more than a prelude to the Emil's sternest test. It may be argued that whereas the entire operational structure of the *Luftwaffe* was designed for the support of ground forces in *Blitzkrieg* tactics, it neither possessed the strategic weapons to defeat the British island nation, nor was it equipped to destroy a relatively sophisticated air defence system that lay over the sea. Yet in the Messerschmitt Bf109E lay the one weapon which, were it to be employed wisely, could destroy RAF Fighter Command as the necessary pre-requisite for a sea-borne invasion. The irony of the situation was that numerous *Luftwaffe* fighter pilots were well aware of their potential superiority yet were over-ruled by a German High Command hell-bent on the protection of their bomber forces. Thus, from the German viewpoint, the over-employment of the Emil in the bomber-escort role not only exposed the fighter to wholly unprofitable combat with British fighters, but wasted the all-important summer days in fruitless attacks on unimportant targets, and added to the exhaustion of frustrated Emil pilots. It was in the critical phase between 26 August and 6 September when Emils were given free rein to sweep the skies of Southern England – relatively unfettered by the bombers – that the RAF came closest to defeat.

Bf109E-1 c. 1940, France.

Bf109E, 7./JG 52, Channel Coast during the Battle of Britain.

Yet even that is to beg the question that superb though the Emil undoubtedly was, it was no more than a relatively short range fighter, capable of snap sweeps over the southern counties of England when based on the Channel Coast. Most units were based in the *Pas de Calais* and their aircraft were able to penetrate only over the counties of Kent, Surrey, Sussex, Middlesex and part of Essex. When committed to sustained air combat they were in great peril of running short of fuel and, more often than not, their pilots were obliged to break off combat first, in the hope that sufficient fuel remained to get them back to the French coast.

When the Battle of Britain opened at the beginning of July 1940 *Luftflotte 2* and *3* fielded about 700 Emils distributed in the same *Geschwaderen* that had fought in the Battle of France, although the number of *Gruppen* had increased to 20 (and further to 22 by 20 July). A significant lesson had been learned in the skies over Dunkirk and had been voiced by influential fighter leaders in the *Jagdgeschwaderen*. The hub-firing FF cannon of the Bf109E-3 was of little value in fighter-versus-fighter combat and, as the destruction of RAF Fighter Command was the stated aim of the *Luftwaffe*, it was decided to phase out the E-3 and replace it with the E-1 and the new E-4.

The latter had entered production early in 1940, but it is thought to have been too late to see any action in the Battle of France. In this variant the hub cannon was removed, the nose-decking guns retained, and the wing machine-guns replaced by MG FF 20mm cannon. It is believed that the first to receive the E-4 was *Jagdgeschwader 27*, one of whose aircraft was damaged in combat with a Hurricane as early as 4 July. It is estimated that by the end of July, 150 E-3s had to be replaced by E-1s and E-4s to have their hub cannon removed. By mid-August the E-1 was the most widely-used variant, but E-4s were fast increasing in numbers. The latter, whose increased armour protection for the pilot gave a

Bf109E-1, France 1940.

slightly better field-of-view, was without doubt the most popular of all Emils, and was thus selected for further development.

The first sub-variant of the E-4 was the E-4/B, a fighter-bomber equipped with belly-shackles to accommodate a single SC250 (550lb) bomb, and the first examples of this version were delivered simultaneously to *Lehrgeschwader 2* and *3 Staffel/Erprobungsgruppe 210*. The latter unit was a crack pathfinder force employed throughout the Battle of Britain to attack the most difficult targets. It had already been using a special version of the E-1 equipped to carry four 110lb bombs in steep-dive attacks. But the arrival of the E-4/B prompted a change in tactics-approach at low level with cannon firing before dropping the bomb and pulling up to provide top cover for other bombers.

Although no accurate figures exist of the number of RAF fighters shot down by Emils during the Battle of Britain, those that do survive enable certain extrapolations to be made. These suggest that about seventy percent of all Hurricanes lost, and sixty percent of all Spitfires lost, as well as ninety-five percent of all Defiants, were due to the Emil.

Thus for a total loss in the air of 587 Emils, during the four months of the Battle of Britain, the German single-seat fighters destroyed about 380 Hurricanes, 210 Spitfires, 18 Defiants and 27 Blenheims – a total of 635 British fighters. The remaining 300-odd British fighters fell to the guns of all other German aircraft types. When one realises that for at least sixty percent of the Battle of Britain, the *Jagdgeschwaderen* were misemployed on bomber escort duties. It is not unreasonable to suppose that had the Emils been entirely free to seek and destroy the opposing fighters throughout the entire battle, RAF losses cannot have failed to have been at least 200 more than they were. Such a loss, sustained by the end of the first week in September, would unquestionably have resulted in the effective collapse of Fighter Command before Göring even had a chance to commit his other fatal mistake, on 7 September, of switching his attack against London itself. That is the measure of the quality of the Messerschmitt Bf109E as a weapon in the Battle of Britain.

Not content with the misappropriation of his fighter reserves during the first three months of the battle, Göring sanctioned the use of one *Staffel* from each *Jagdgeschwader* in the role of nuisance bombers. Burdened with a single SC250 (550lb) bomb on the belly rack, the Emil was now employed in flying over south-eastern England at high level and dropping the bomb, virtually at random, on the prominent conurbations. Such sorties were in fact a tacit admission by the German commander-in-chief that his conventional bomber forces were unable to bring about a decisive conclusion to the daylight battle over Britain.

The Emils employed in this final phase of the Battle of Britain were E-4/Bs (equivalent to the earlier E-1/B) and the E-7, the latter having started to appear in small numbers on the Channel Coast towards the end of August. It was basically an E-4 with modified fuel system to allow carriage of a long range belly fuel tank, but there is no record of these tanks being used during the battle, rather that the shackles were used to carry the bomb. The first such aircraft to be shot down was from I/LG 2, destroyed on 31 August by a Hurricane flown by Plt Off H.A.C. Bird-Wilson of No. 17 Squadron over Maidstone. The German pilot, *Oberleutnant* von Perthes, was killed. On this day *Luftwaffe* records showed twenty-seven E-7s at operational strength.

Oblt Heinz Ebeling flew this Bf109E of 9./JG 26 'Schlageter'. Twelfth victory bar being applied.

Bf109E, II./JG 51 'Mölders', shot down in Battle of Britain.

Three other E-series 109s reached operational units of the *Luftwaffe* before the end of the Battle of Britain. They were the E-5, E-6 and E-8. The first two were fighter-reconnaissance variants with all wing armament removed and a camera installed in the fuselage behind the pilot's seat. They differed in that the E-5 retained the DB601A whereas the E-6 was powered by the DB601N – a new variant of the engine employing improved fuel injection and automatic hydraulic supercharger drive coupling. Some E-4s were retrospectively fitted with the DB601N engine thereby producing the E-4N sub-variant. The E-8 was the designation used to embrace Emils which incorporated all features developed to date, including improved cockpit armour, universal fuel system, provision for belly rack pick-up, provision for wing cannons and hub-firing MG FF gun, and was powered by a 1,200hp DB601E engine.

Standard E-3s were taken from the production line, fitted with wings of increased area and upper-surface spoilers; rear fuselage arrester hooks were fitted and the outer wing panels were designed to fold – this operation being undertaken manually. After flight trials of the first aircraft, the contract was extended and, by the time work on the *Graf Zeppelin* was suspended, about fifty aircraft had been completed. Such was the extent of the modifications that nothing useful could be achieved in converting the aircraft back to 'standard', and most of the 109Ts were delivered to I/JG77 stationed in North Germany early in 1941.

The Messerschmitt Bf109 proved notably suitable for continued development and, although the classic Emil gradually disappeared from front-line squadrons in 1941, its place was progressively taken by the 'F', 'G' and 'K' series which continued in operational service right up to the end of the Second World War. Yet many of the great Emil pilots were reluctant to part with the older type when the 'F' series appeared. Walter Oesau, a leading pilot of the Spanish Civil War and the Battle of Britain, simply refused to transfer to the newer type so long as spares existed to keep his Emil in the air.

Bf109E-1

Leading particulars

Wing span: 32ft 4.5in, area 176.53sqft. *Length:* 28ft 4in. *Height:* 8ft 2.5in. *Weights:* tare 4,056lb, gross 4,431lb, max permissible 5,523lb. *Max speed:* 289mph @ S/L, 302mph @ 3,280ft, 334mph @ 19,685ft, cruise 298mph @ 13,120ft. *Ceiling:* 34,450ft. *Rate of climb:* 3,050ft/min @ S/L. *Height to time:* 16,400ft in 6.2mins. *Range:* 410 miles @ 205mph @ 3280ft. *Engine:* Daimler Benz DB601A of 1,050hp @ T/O, 1,100hp @ 12,140ft. *Armament:* two 20mm MG FF cannon with 60rpg in wings and two 7.9 MG 17 m/gns with 1,000 rpg in wings.

Bf109E-3

Leading particulars

Dimensions as for E-1. *Weights:* tare 4,189lb, gross 4,685lb, max permissible 5,875lb. *Max speed:* 293mph @ S/L, 336mph @ 19,685ft, cruise 300mhp @ 13,120ft. *Ceiling:* 34,450ft. *Rate of climb:* 3,280ft/min @ S/L. *Height to time:* 19,685ft in 7.1mins. *Range:* 410 miles @ 233mph. *Engine:* Daimler Benz DB601Aa of 1,175hp @ T/O, 1,020hp @ 14,765ft. *Armament:* as E-1 or one engine mounted 20mm MG FF/M cannon with 200 rounds.

Bf109E-4

Leading particulars

Wing span: 32ft 4.5in, area 174sqft. *Length:* 28ft 8in. *Height:* (tail down over propeller) 11ft 2in. *Weights:* tare 4,440lb, gross 5,520lb. *Max speed:* 357mph @ 12,300ft, cruise 298mph @ 62.5% rated power, landing 78mph. *Ceiling:* 36,000ft. *Rate of climb:* 3,100ft/min @ S/L. *Range:* 412 miles @ cruise speed and power. *Engine:* Daimler Benz DB601Aa of 1,150hp @ 2,400rpm @ T/O. *Fuel:* 88gals in fuselage under pilot seat. *Propeller:* VDM electric CP/FF 3-blade metal. *Armament:* two 7.9 MG 17 m/gns with 1,000rpg in upper nose decking and firing through troughs. Two 20mm MG FF cannon with 60rpg in wings.

Bf109E-7
Leading particulars
Wing span: 32ft 4.5in, area 174sqft. *Length:* 28ft.8in. *Height:* 11ft 2in. *Weights:* tare 4,440lbs, gross 6,100lbs *Max speed:* 359mph @ 12,300ft. *Ceiling:* 36,500ft. *Rate of climb:* 3,300ft/min @ S/L. *Range:* 680 miles. *Engine:* Daimler Benz DB 1,200hp. *Armament:* One 200mm MG FFM hub mounted cannon and four 7.9mm MG 17 m/gns in nose and wings.

Structure E-Models

Fuselage of oval section is metal monocoque manufactured in two sections and joined longitudinally top and bottom. Longitudinal stringers and vertical panels. Flush riveted overall. Wings, low cantilever monoplane. All metal, single spar with metal stressed skin flush riveted. Three fuselage attachment points on spar flanges. Entire trailing edge with hinged, slotted ailerons outboard and slotted flaps inboard. Handley-Page type autoslats on outboard leading edges. Tailplane mounted on cantilever fin and braced to fuselage by single strut each side. Balanced rudder and elevators. Metal frame with moved surfaces covered in fabric. Undercarriage retracted upward and out. Fixed tail-wheel.

Bf109E-7 of II Schlachtgruppe, Lehrgeschwader 2, St Omer, France, 1940.

Bf109E, I/JG 2 'Richthofen', France, 1940. Yellow cowling and rudder.

Bf109E of Major *Helmut Wick,* Geschwader Kommodore, Jadgdeschwader 2 'Richthofen'.

Bf109E, JG 26 'Schlageter', *North France. Blanked spinner hub.*

Bf109Es of I./JG 3 'Udet', *Pas de Cailas, 1940 Battle of Britain insignia.*

Bf109E, I./JG 2, France, summer 1940. Yellow cowl and rudder.

Bf109E-4/B fighter bomber. 3 Staffel, Gr. 210. Carries 250lb bomb.

Bf109E-7, II (Schlacht/LG 2), St Omer. Armed with four 110lb bombs. Battle of Britain period. Yellow cowl and rudder.

Bf109E-1, JG 53, Belgium, May 1940.

Bf109E-1. I Gruppe, Jagdgeschwader 77 *Belgium during Battle of Britain. Transferred from* Loftflotte 5, *Norway.*

CHAPTER FIVE

On Other Fronts

After the daylight phase of the Battle of Britain the German fighter strength on the Channel Coast was slowly diminished by the withdrawal of five *Jagdgeschwaderen*, until by mid-1941 only two, including JG26 'Schlageter', remained. They were in fact being re-equipped by the new Bf109F. By the end of 1940 a total of 2,408 Bf109s had been built, and Germany had found it possible to send 284 of them to countries friendly to the Central Powers. In addition to these premeditated deliveries, half a dozen others had landed intact in England and in neutral territory.

Switzerland was the largest customer for the Emil. Thirty E-1s were delivered in 1939, armed with two 20mm Hispano and two 7.45mm guns. Fifty more were ordered and deliveries of these were completed on 27 April 1940. Within three months Nos 6, 7, 8, 9, 15 and 21 *Fliegerkompagnien* of the Swiss Air Force had achieved operational status and were having their work cut out to guard their airspace from incursions by wandering *Luftwaffe* aircraft. Imported Emils were registered from J-311 to J-390. Licence production was also undertaken by the Dornier-Werke AG of Alterrhein, but only nine home built examples were completed (J-391 to J-399), apart from a quantity of spare parts.

Jugoslavia also negotiated for licenced production of the Emil, following the process of reorganisation that commenced in her Air Force in 1938.

In the event licence production never got under way; and an initial order for fifty Emils to be supplied by Germany was completed in 1939 and 1940, when a second order for fifty was placed. However, by the time German forces marched into Jugoslavia on 6 April 1941 only seventy-three aircraft had been delivered to the *Jugoslovensko kraljevsko ratno vazduhoplovstvo*, and on that day only forty-six were serviceable. These were deployed with the 2nd and 6th Fighter Regiments and with the Fighter Training School (SELS). Frequently required to fight against similar aircraft of the *Luftwaffe*, the Jugoslav pilots acquitted themselves gallantly but were soon overwhelmed.

Romania's joining with Germany in the Tripartite Pact on 23 November was followed by orders for a quantity of German aircraft, including forty Bf109E-4s. However, these aircraft had not been operationally deployed with the *Fortelor Aeriene Regal ale Romania* by the time the FFAR joined elements of *Luftflotte 4* in the German assault on Russia in 1941. In 1942 the Romanian elements were withdrawn to re-equip, and the aged Polish PZL P.24Es were replaced by sixty-nine Bf109E-4s newly arrived from Germany, being deployed as equipment of two fighter Groups of the 1st Air Corps. These aircraft fought for about six months in the Ukraine before being replaced by Bf109Gs. Hungarian elements of *Luftflotte 4* also received forty Bf109E-4s as replacements for their Fiat CR42 biplanes which had suffered severely on the Russian Front.

Bf109. Note wing/fuselage national markings. Possibly Combined Insurgent Force, Hungary.

Bf109E-4 Slovakian Air Force.

Bulgaria, despite her membership of the Tripartite Pact, took little part in the early Russian campaigns. Her Air Force, then an integral part of the army, did take delivery of nineteen Bf109E-4s, but it is not likely that these saw combat service. On the other hand, two squadrons of the semi-autonomous Slovak Air Force, equipped with sixteen Bf109E-3s, fought alongside the *Luftwaffe* during the invasion of Russian in 1941. Only one other nation specifically ordered Emils; this was Japan, who took delivery of two Emils (probably in 1940) in preparation for licence production of the type by Kawasaki, but this failed to materialise.

The Desert War

Returning to the Emil in service with the *Luftwaffe*, failure by Italy to defeat the British forces in North Africa led to the despatch of German forces in ever growing strength early in 1941. Up to this point the RAF had enjoyed undisputed air superiority over the desert, despite its fighter squadrons still being equipped with very early versions of the Hurricane I and obsolete Gladiator biplanes. The tide turned dramatically with the arrival of two *Jagdgeschwaderen*, JG27 and JG53 'Pik As', equipped with tropicalised versions of the Bf109E-4/N – distinguishable by their lengthened air intake fairing incorporating sand filter on the port side of the nose. Some fighter-bombers soon followed, as did tropicalised E-7s; as added protection from ground fire, extra armour was attached under the wing radiators and engine, this version being designated the E-7/U2.

Bf109E-4/Trop, Western Desert. The shape is difficult to spot against background.

Bf109E-4 (N) Trop, Libya.

Not surprisingly the appearance of Emils in Africa and over the Mediterranean quickly brought an end to Gladiator operations. It was not long before Spitfires were being sent out from Britain as soon as they could be spared, but the Hurricane continued to serve over the desert for many months and suffered accordingly at the hands of the German fighter. Supreme among the German pilots during 1941 and 1942 was the famous Hans-Joachim Marseille who, with JG27, flew the Bf109 from the time of his arrival in the Mediterranean theatre in April 1941 until his death on 30 September 1942. In the course of those eighteen months he shot down no fewer than 151 RAF aircraft, and on one famous occasion (1 September 1942), in the course of three sorties, shot down 17 British fighters, of which it is said that 8 fell in ten minutes!

The Emil survived in service in North Africa rather longer than on other German fronts. In the Balkan campaigns of 1941, III/JG77 continued to fly E-4s, but by the time Hitler attacked Russia in June that year almost all the Emils on JG1, 3, 51, 52, 53, 54, and 77 had been replaced by the F-1 and F-2. Some E-4s are known to have survived on JG54 and SG1. The only other variants, which appeared in 1941, were the E-7/Z and the E-9. The former was fitted with the GM 1 boost which employed injection of nitrous oxide into the supercharger to reduce detonation.

The E-9, powered by a DB 601E, was a reconnaissance version otherwise similar to the E-5 and E-6; without wing armament, they were fitted with an Rb 50/30 camera in the fuselage, and flew with a ventral 66-gallon long-range fuel tank.

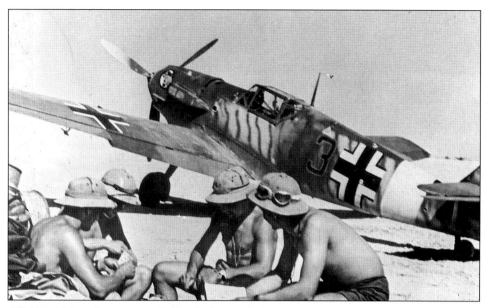

Bf209E-4/Trop of I./JG 27. Pale blue fuselage with dark green stripes.

Be 109T-2

One further variant should be mentioned, the Bf109T (T standing for *Träger*, or carrier). In 1940 work was still progressing apace on Germany's new aircraft carrier *Graf Zeppelin*. Accordingly Fieseler was instructed to prepare ten converted Bf109s as shipborne fighters.

It was Admiral Raeder who proposed this view of challenging British sea power, not only with the accepted view of U-boats, but also a surface fleet which would have the so called 'Pocket Battleships' and other units, but also aircraft carriers. He recognised that a combination of surface fleet and U-boat operations could shorten the war by cutting England off from it vital overseas supplies from America in particular, the Commonwealth and other neutral countries.

Wide span version of the Bf109T-2 after conversion as FB from carrier design.

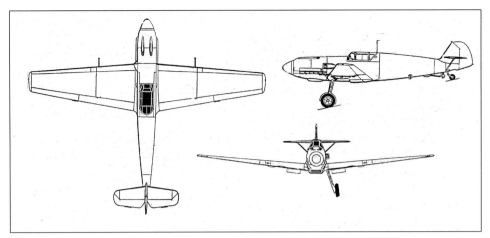

The Bf109T-1 carrier borne fighter intended for the German Navy aircraft carrier Graf Zeppelin. *It was never used as the carrier was never completed, and all production aircraft were converted to the T-2 standards as fighter bombers.*

However, ships take a long time to reach active service from the laying of the keel to finished boat, and the *Luftwaffe* Chiefs put forward the persuasive argument that a combination of long range aircraft and U-boats would accomplish more.

This did not prevent two carriers being laid down, the *Graf Zeppelin* and *Peter Strasser*. In conjunction with this the Carrier Group (*Trägergruppe*) was formed with Bf109B fighters and Junkers Ju87 dive-bombers. The Messerschmitt company had been instructed to develop a specialised version of its landplane variant, the 109E and they produced a design designated as Bf109T (*Träger*) in early 1939.

The 109E's wing had been increased in area by two feet wide outer panels complete with slots and ailerons. Catapult points were installed between the fifth and sixth fuselage frames, and an arrester hook attached aft of the seventh. Local structure at these points was strengthened. Manual wing folding was accomplished by a hinge in the spar immediately outboard of the gun bays. This reduced folded width to 13ft 4in.

Armament was specified as two 7.9 MG 17 m/gns in the fuselage with alternative wing mounted guns such a MG 17s and MG FF canon. The proposals were accepted and all development transferred to the Fiesler Works. Ten production 109E-1s were to be converted initially as pre-production Bf109T-0s. A contract for sixty production 109T-1s followed.

It had been intended that these production aircraft be supplied to II/JG186 for evaluation, but events overtook the scheme and all work on the main element *Graf Zeppelin* was suspended during October 1939. Hitler had sensed he could win the war with the *Blitzkreig* tactics on land and did not want to divert effort into the carrier force. The unit selected for the trials of the fighter became III/JG77, and production of the sixty 109Ts was halted.

Fiesler was then instructed to complete the 109T-1s but without carrier equipment and substitute a ventral fuselage rack for carriage of a 66gal fuel tank, four 110 or one 551lb bombs to allow operation from short airstrips or even small fields. The new variant was the Bf109T-2. Although work did start again on the *Graf Zeppelin* the 109T-2s were not considered suitable. Survivors were, however, still flying in 1944.

Bf109T-2

Leading particulars

Wing span: 36ft 4in, area 188.368sqft. *Length:* 28ft 9in. *Height:* 8ft 6.5in. *Weights:* tare 4,409lb, gross 6,173lb, max permissible 6,786lb. *Max speed:* 304mph @ S/L, 328mph @ 6,560ft, 357mph @ 19,685ft, cruise 272mph @ S/L, 343mph @ 16,400. *Ceiling:* 34,450ft. *Rate of climb:* 3,346ft/min @ S/L. *Height to time:* 26,250ft in 10mins. *Range:* 454 miles @ 283mph @ 29,530ft. *Engine:* Daimler Benz DB 601N of 1,200hp @ T/O, 1,270hp @ 16,400ft.

Bf109E-4 (N) Trop, I./JG 27.

Bf109E-8 was flight tested on fixed ski undercarriage, 1941-42.

Bf109E-4/B 'Grünherz', *Russia 1942.*

Bf109E-4/Trop of I./JG 27 over the desert.

Well known photo of Bf109E-4(N) Trop of I./JG 27.

Bf109E-4 (N) Trop, I./JG 27 at take off.

Bf109E-7, 7./JG 26 'Schlageter', Sirte, Libya. No sand filter. Note ventral fuel drop tank.

Bf109E-7, 7./JG 26, Libya, 1941.

Bf109E-4/N, 7./JG 26 'Schlageter', Sirte, Libya. Joachim Müncheberg.

Bf109E-4(N) Trop, JG 26, North Africa, 1941.

Bf109E of JG 26 'Schlageter', *Sirte, Libya.*

Bf109E, I./JG 27, North Africa, 1941. Early type sand filter.

Bf109E-4/N, I Gruppe, Jadgeschwader 27, *Tripoli, North Africa. Note drop tank and early filter.*

Bf109E-4/B, JG(S) LG 2, Eastern Front, Russia 1942. Four bombs under fuselage.

Bf109E-4/B, 8./JG/SKG 210, El Daba.

Bf109E 'Emil', of I./JG 27, North Africa, 1941.

Bf109E of I./JG 27.

Bf109E of I./JG 27.

Bf109E-4/N flown by Oberefeldwebel *Albert Esplenlaub of I/JG 27, Tripoli, Libya, 1941. Fourteen confirmed kills.*

Pilot of Bf109F-7, 7./JG 26.

Bf109E-4 refuels on desert airstrip.

Bf109E-3, 1941.

Bf109Es and Ju87s in Russia. I./JG 77.

Bf109E-4, II./JG 54, Russia, autumn 1941.

Bf109Es.

Bf109E during u/c retracting tests.

Refill for auxiliary tank of this Bf109E-4 of JG 26 Schlageter.

An E-1 accelerates for take off in the desert. Note the African lion head motif on engine nacelle of JG27.

CHAPTER SIX

Bf109F Series

After the Battle of Britain the *Luftwaffe* wanted either a new fighter or improved versions of the 109E if it was to continue to keep ahead in the air war. Messerschmitt had already anticipated this need and improved designs had been firmed and put into production. The company had received feed-back from the battle which revealed weaknesses in the fighting aircraft that could be resolved by either rapid modification or constant updating of original design.

The 109E had served the *Luftwaffe* well in Poland and in the first few months of 1940 but the RLM now wanted an improved fighter. A new, more powerful version of the Daimler Benz engine promising 1,350hp at take-off had been developed and this, together with other modifications to the flaps, leading edge slots and ailerons produced a reduction in wing span. There were changes also to the engine cowling when the original angular shape was streamlined. The rudder was decreased in areas and a cantilever tailplane led to deletion of the struts.

Work began on four prototypes and ten pre-production Bf109F-0 aircraft in March 1940. The prototypes were the V21 which incorporated all modifications except for the engine; V22 had the new DB 601E powerplant; V23 was flight tested with detachable wing tips and the V24 had minor changes to the supercharger intake and deeper oil-cooler fairing. Armament was four wing-mounted machine guns and a single 15 or 20mm MG 151 hub mounted cannon. The first flight of a prototype took place on 10 July 1940, too late for any to be used in the Battle of Britain. The second prototype 109 V22 had a pre-production DB 601E engine as did V23 and V24 but with modifications.

Also, when production did get under way the improved Daimler Benz engine was not available in quantity and early 109F models had to use the DB 601N engine. Standard production 109F-1s with the new engine became available in November 1940, but problems soon became evident as engine vibration could lead to loss of control. Investigation finally led to strengthening of the tail assembly which was vibrating during full power. When the first production aircraft were flight tested it was found that the engine was vibrating badly with the result that two aircraft crashed.

The first units to operate the F version were *JG2 'Richthofen'* and *JG26 'Schlageter'*. In spring 1941 Fighter Command's Spitfire I had difficulty intercepting the new 109F as it flew, and could fight, at a higher altitude. Supermarine and the Air Ministry were already working on the Spitfire replacement, the Mk III with the new Merlin 60 engines, but the 109F had to be challenged. It was this challenge that led to the compromise Spitfire, the Mk V with Merlin 45 engines. The Spitfire Mk V had improved climb and dive capabilities and was certainly on a par with the Bf109F.

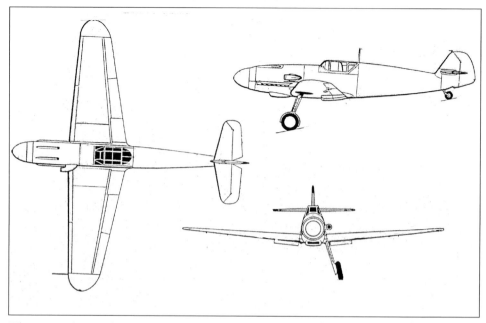

The Messerschmitt Bf109F Series covering F-1 to F-6 variants.

Bf109 V23 prototype as modified with the Me309 trials. Tri-cycle u/c.

Bf109F with redesigned nose, France.

A total of 440 Bf109Fs formed the major proportion of the *Luftwaffe* squadrons that took part in Operation *Barbarossa* on 22 June 1941 and it is claimed that they shot down large numbers of Soviet aircraft (300 plus) on the first day of the invasion. Victories did come easy to the Messerschmitt pilots as the number of kills recorded, both on the ground and in the air were 800 and 400 respectively. However, the *Luftwaffe* also lost fighters and the wastage eventually outpaced replacements.

To enable the 109's engine to achieve higher emergency power output a number had GM-1 nitrous-oxide injection, being designated as the Bf109F-2/Z. Meanwhile the desert war had expanded and, like the Allied fighters, the 109F required a sand filter. This produced the Bf109F-2/Trop which began to reach the *Luftwaffe* in Libya to II./JG27. The Bf109F-3 was fitted with the DB 601E engine now using 87 octane B4 fuel.

Weight escalation was reducing performance of the 109 and when the Bf109F-4 was available it had the 20mm MG 151 installed, with ammunition limited to 150 rounds. Self-sealing fuel tanks, an armoured windscreen and pilot armour added weight and the only solution to performance was a more powerful engine. The F-4/B was developed as a replacement for the Junkers Ju87 Stuka, which was now totally outclassed in European theatres. The Messerschmitt fighter was used for anti-shipping and coastal attack over Southern England, a role once carried out by the Stuka. The E-104/B could carry a ventral fuselage rack and one 551lb bomb, and this combination kept Fighter Command alert. The tropical version was the 109F-4/Trop.

Additional armament changes consisting of two 20mm MG 151 cannon with 120rpg in detachable gondolas under the wings for use against the RAF's bomber squadrons produced the Bf109F-4/R1. A number of aircraft had a ventral fuselage rack carrying a single 551lb bomb or a 66gal drop fuel tank and provision for four, 110lb bombs and were

Bf109F-0 pre-production aircraft model, one of ten.

called the 109F-4/R6. The 109F-4/Z had GM 1 equipment, and the 109F-5 with the hub mounted cannon deleted and vertical camera in rear fuselage provided a fighter reconnaissance version along with the 109F-6 with no armament and interchangeable camera installations.

As with all production aeroplanes the 109F was used for experiment including an F-2 with four EG 65, 73mm Rheinmetal-Borsig RZ rockets and launchers, for use as a weapon against the RAF bomber formations. A F-4 was fitted with a butterfly tail unit and the Bf109 V24 prototype was tested with the Me309's ventral radiator bath. The V23, also used for Me209 trials, had an awkward looking tricycle undercarriage installed.

The Bf109 V31 prototype had a wide, inward retracting, main wheel undercarriage, modified radiator bath. V30 and V30a flight tested the Me309's pressure system. Performance of the impressive Fw190 fighter with the BMW air-cooled radial engine led to a 109 being used as a flying test with this unit. A second had a Jumo engine fitted with an annular radiator in the form of the Ju88 bomber installation.

Bf109F-2

Leading particulars

Dimensions: as for F-4. *Weights:* tare 5,188lb, gross 6,393lb, max permissible 6,872lb. *Max speed:* 321mph @ S/L, 388mph @ 21,235ft, cruise 310mph @ S/L. *Ceiling:* 36,090ft. *Rate of climb:* 3,860ft/min @ S/L. *Height to time:* 16,400ft in 2.6mins. *Range:* (with 66gal drop tanks) 442 miles @ 314mph @ 16,400ft, 528 miles @ 298mph @ S/L. *Engine:* Daimler Benz DB 601N of 1,200hp @ T/O, 1,270hp @ 16,400ft. *Armament:* one 15mm MG 151 cannon with 200ropg and two 7.9 MG 17 m/gns with 500rpg.

Prototype Bf109 V31 test bed for the retractable radiator bath and wide track, inward moving u/c for the Me309.

It was the Russian theatre that once again provided the major scoring 'Aces'. *Major* G. Gordon Gollob 100, Hermann Graf 104, *Hauptmann* Heinz Bär 100.

Final use for the F-3 was the Bf109F and Junkers Ju88 *Beethoven-Gerat (Mistel)* pick-a-back concept when a fighter (109 or Fw190) would be attached to the 88's dorsal fuselage, flown to a target with the fighter detaching and the fighter returning. At least, that was the theory. All work on the project was obscure, but the bomb had a head that blasted a stream of hot metal into the target that destroyed everything in it path.

The 109F was supplied to Italy providing equipment for two *Regia Aeronautica Gruppe*. The Hungarian Air Force had the *1/1 Fighter Squadron* attached to a *Jadgeschwader II./HG 51* in the Stalingrad area of fighting.

Bf109F-3
Leading particulars
Dimensions: as for F-4. *Weights:* tare 4,330lb, gross 6,054lb, max permissible 6,872lb. *Max speed:* 331mph @ S/L, 390mph @ 21,000ft, cruise 310mph @ S/L. *Ceiling:* 37,090ft. *Rate of climb:* 3,860ft/min @ S/L. *Height to time:* 16,400ft in 2.6mins. *Range:* (with 66gal drop tanks) 442 miles @ 314mph, @ 16,400ft, 528 miles @ 298mph @ S/L. *Endurance:* 1hr 42mins. *Engine:* Daimler Benz DB 601E of 1,300hp @ T/O 1,270hp @ 16,400ft. *Armament:* One 15mm MG 151 cannon with 200rpg and two 7.9 MG 17 m/gns with 500rpg in upper cowling.

Bf109F-4
Leading particulars
Wing span: 32ft 5.75in, area 174.376sqft. *Length:* 20ft 0in. *Height:* 8ft 6in. *Weights:* tare 5,269lbs, gross 6,393lb, max permissible 6,872lb. *Max speed:* 344mph @ S/L, 346mph @ 9,840ft, 388mph @ 21,685, cruise 310mph @ S/L, 355mph @ 16,400ft. *Ceiling:* 39,370ft. *Rate of climb:* 4,290 ft/min @ S/L. *Height to time:* 16,400ft in 5.2mins. *Range:* 528 miles @ 298mph @ S/L. *Engine:* Daimler Benz DB 601E-1 of 1,350hp @ T/O, 1,300 @ 18,000ft. *Armament:* One 20mm MG 151 cannon with 150 rounds and two 7.9mm Mg 17 m/gns with 500rpg.

Bf109 V24, prototype for the F Series.

Bf109F-4/Trop of II./JG 53.

Bf109F of II./JG 27. Clear view of tropical filter on this captured aircraft.

Bf109F-2

A Bf104F can be seen with the twin MG 17s in nose. Cannon shells loaded into wing.

Underside view of Bf109F-4/R1 as it approaches to Tunisian airfield. No filter, MF 151 wing cannon in gondolas. Yellow wing tips. Spring 1943.

Technical officer checks cannon barrel a Bf109F-2 of II./JG 54, Tunisia, 1942.

Bf109F-4.

Pair of Bf109Fs.

Bf109F-4, flown by Gruppen Kommandeur III./JG 2.

Bf109F shot down in desert, May 1943.

Marseille and his 109F.

Bf109F, Quotofina, North Africa, 30 September 1942.

Bf109F-1. III./JG 3 'Udet', Russia, 1941.

Close up of Bf109F-2 which was captured after landing near Dover, 10 July 1941.

Bf109F-5 I/(F) 122.

Bf109F-5s of 4(F) 124.

Bf109F of II./JG 54, Russia, winter 1942.

Bf109F, 7./JG 54, III Gruppe, Russia.

Bf109F-2, II./JG 54. Interesting scheme.

Bf109F-2, 7./JG 54, Siwerskaja, Russia.

Bf109F-2, Russia. Note large cowling panels.

Bf109F-2.

Bf109Fs of JG 53 'Pik As', desert 1942.

Bf109F-2 of II./JG 27 refuelling on Sicilian airfield.

Bf109F-2 of II./JG 27, Sicily.

Bf109Fs.

Undercarriage testing on a Bf109F, desert airstrip. Note engine at high revs and human tail ballast.

Bf109F in desert setting following engine start.

Bf109F-2 of II./JG 27 is manhandled.

Bf109F-2 undergoes maintenance.

Bf109F-4 in winter camouflage, Russia 1943.

Bf109F-4/Trop takes on fuel in desert.

Bf109F-4 in Russia in winter.

Three pilots of JG54. Gunther Fink, Staffelkapitän of 8./JG54, Rudolf Klemm, as an Oberfeldwebel with III./JG54, and Heinrich Jung, Staffelkapitän of 4./JG54. Klemm had just been credited with his first victory. He went on to become a Major and Kommandeur of IV./JG54. His Bf109F is seen in the background.

Bf109F of I./JG 27 North Africa. Note underwing gun gondolas.

Air to ground view of Bf109F shows compact size of aircraft.

Battle damaged Bf109F.

Bf109F approaches to land.

Exposed Daimler Benz DB 601 engine in Bf109F.

CHAPTER SEVEN

The Bf109G 'Gustav' Series.

Progressive development of the Spitfire was accelerated by the advances made in the equipment of the *Luftwaffe*, in particular the Bf109. When the 109F made its first sorties over Southern England in October 1940, the RAF was still awaiting the successor of the Mk I/II Spitfire. As the Mark III, it would have been powered by the Merlin XX engine and, eventually, the Merlin 61 Series. But, the new variant of the Messerschmitt fighter resulted in a crash programme by Supermarine and Rolls-Royce, and a combination of a strengthened airframe, together with a modified Merlin XX, the Merlin 45, resulted in the Spitfire Mk V which could meet the Bf109F on equal terms.

As the war progressed and Germany came under pressure on two fronts when Russia was invaded, the pace of progress slowed. Production was the main demand and the Bf109F series was soon to be on a par with the latest Russian fighters as new Soviet models appeared. A new Mark, or design of aeroplane was demanded and rather than embark on a long term programme the Messerschmitt team reworked the basic 109F airframe, first by adding a pressure cabin to enable it to operate at a higher altitude, and a more powerful Daimler Benz engine, the 605A, which provided Messerschmitt's fighter with a take off power of 1,475hp. The armament of one 20mm MG 151/20 cannon and two 7.9mm MG 17 machine guns was retained and there was provision for nitrous oxide for the emergency burst of power when required. The penalty for all these, and other, modifications, was a gross weight which had increased to 7,055lb.

Produced in larger numbers than all other previous or, indeed, future models of the Messerschmitt Bf109, the G-series, particularly the more heavily armed variants, was considered by many fighter pilots of the *Luftwaffe* to be inferior to its immediate predecessor, the Bf109F. The Bf109G, christened 'Gustav' by the *Luftwaffe*, entered service in the summer of 1942, and soon regained the ascendancy the earlier Mark had possessed over the Spitfire. The *Luftwaffe* now possessed two potent warplanes, and the Royal Air Force could do little to effect the balance of air power until the Spitfire Mark IX reached Fighter Command squadrons.

Design of the early production Bf109G's followed closely that of the Bf109F series, and could only be distinguished by the of the Daimler Benz DB 605A engine in place of the 1,300hp DB 601E. A dozen pre-production Bf109G-0s were built to Bf109G standards, but they lacked the DB 605 engine which was being rushed into production and there was insufficient numbers with which to fit into the new Mark. As a result only a small number of early production 109Gs were powered by the DB 601E, and armament included machine guns and one engine-mounted cannon. The BF 109G-0 also had structural strengthening.

The first production model that followed the Bf109G-0 was the G-1, essentially similar to the G-0, but now powered by the 1,475hp DB 606A-1 engine plus GM-1 power

Bf109G-6/R4 with longer tailwheel strut.

boosting. This was a system which used nitrous-oxide injected into the supercharger to boost power above the engine's rated altitude. A tropical version, the BF 109G-1/Trop, had tropical filters and two 13mm MG 131 machine guns in place of the MG 17s. These guns had very large breech blocks that protruded above the engine cowling profile and had to be covered by large fairings, raising the drag factor. Pilots were soon to find a sobriquet – 'Buele' (bump) to add to that of Gustav.

The Bf109G-2 was similar to the G-1 and lacked a pressure cabin, a desirable feature for a high altitude fighter, but the High Command had decided it was of insignificant value operationally. As the higher echelons pointed out, the variant was intended to undertake the fighter-reconnaissance role. As such it was lightly armed and carried just two MG 17 machine guns in the upper engine cowling, but did have provision for two rearward firing guns of similar type which could be mounted in a ventral pack. It was an odd idea, one that had been considered for the Spitfire after a Frenchman had offered a similar scheme to the Royal Air Force.

Inevitably a fighter-bomber variant was available, the Bf109G-2/R1 with provision for a single 551lb SC250 or a 1,100lb SC500 bomb mounted on a ventral fuselage rack. Two standard overload fuel drop tanks were an additional fitment. Because of the minimal ground clearance of this Mark when carrying such an ordnance load, trials were carried out of an auxiliary, jettisionable undercarriage leg which could be fitted before take off to support the rear fuselage, and when the aircraft was airborne the extra oleo was jettisoned. Surprisingly, the scheme was successful, but the demands of quantity production led to it being abandoned and many 2/R1 pilots made bumpy take-offs when fully loaded.

The Bf109G-3 was for all purposes a G-1 but with a FuG 16Z radio in place of FuG 7A. The G-4, an unpressurised version of the G-3, and the Bf109G-5 introduced an installation of the more powerful DB 605D engine which had a larger supercharger and MW-50 power boosting, originally tested on the Bf109G-5. When added to 100 octane fuel and injected into the engine's cylinders, the water-methanol mix increased thermal efficiency and boosted power to a maximum of 1,800hp. An MW-50 pack was carried in a jettisonable under-fuselage tank and could be operated by means of a lever in the cockpit. Standard armament of the BF 109G-5 was adopted as two 13mm MG 131 machine guns and an engine-mounted 20mm MG 151/20 cannon.

Bf109G-2/R1, with auxiliary take-off wheel to allow clearance of SC500.

95

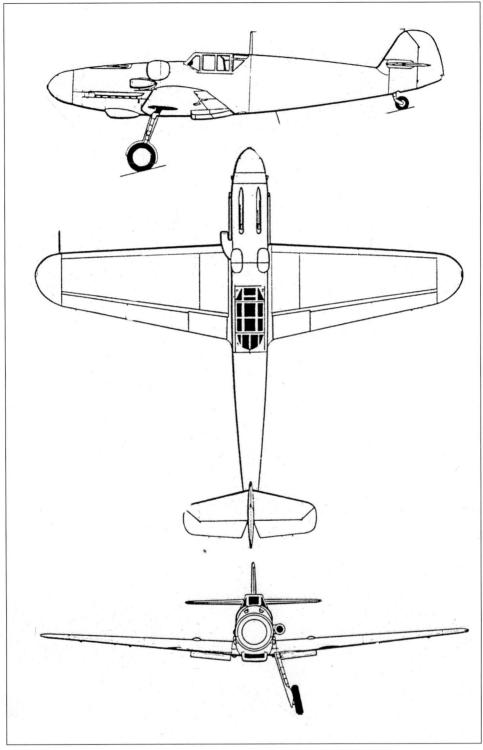

G.A. of the Bf109G-5 with fuselage mounted MG 131 machine guns.

The Bf109G-5/R2 was fitted with the new, taller tail fin and rudder of wooden construction, and a lengthened tailwheel, in an attempt to cure the aeroplane's vicious swing on take-off due to engine torque. The new tail unit was much heavier than the standard duralumin structure, and to bring the centre of gravity back into acceptable limits a weight was bolted under the oil tank bracket.

By far the most numerous model of the 'Gustav' was the Bf109G-6 which could be powered by the DB 605 AM, AS, ASB, ASD or ASM engine. Armament consisted of two MG 131 machine guns, an engine-mounted 30mm MK 108 cannon and two underwing 20mm MG 151/20 cannons. This combination of heavy armament was effective when used against the American daylight bombers, but the effect on its manoeuvrability was almost catastrophic. The Bf109G-6/U4 was an attempt to cure this fault and could carry two 30mm MK 108 cannons underwing in place of the MG 151/20s. The G-6/U4N was fitted with FuG 350 Naxos with a special rotating antenna aft of the cockpit housed under a transparent blister, and was for tracking the H2S radar equipment carried by RAF Pathfinder aircraft.

The G-6R2 carried a 210mm WGr 21 'Dodel' rocket tube under each wing in place of the cannon gondolas. The G-6/R2 Trop was similar, but equipped with tropical filters. The BF 109G-7 was a proposed variant employing all the detailed improvements introduced by the G-6, but was abandoned in favour of the much faster G-10 without reaching the production stage. Produced as a FR version of the 109G-6, the 109G-8 was considered to be a high speed, reconnaissance type with a single MK 108 30mm cannon firing through the propeller boss and could carry either a RB 12.5/7 or RB 32/7 camera in the fuselage. It was serving with the *Luftwaffe* in late 1943 with, among other units, NAGr 13 in France. The unit's main function was to gather intelligence on the Allied build up for the expected invasion across the Channel. A small provisional batch was constructed before sufficient numbers of the G-10 were available.

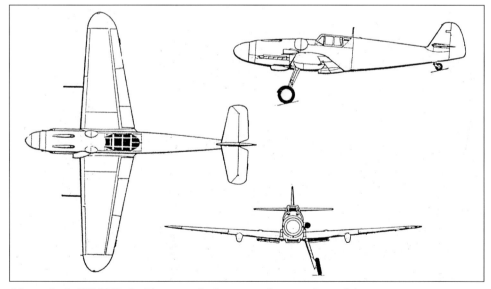

Messerschmitt Bf109G-6 with nose and wing mounted cannon in gondolas.

Bf109G-6/R2 'Pulk Zersteror' armed with Wfr Gr 21 mortars below wings

By far the fastest variant of the 109s was the G-10 whose DB 605D engine, boosted with MW-50 water-methanol injection, was known to reach speeds of 428mph at its service height of 25,000 feet. But, a price had to be paid for that acceleration – a restricted endurance and a range of just 360 miles. Climb was spectacular, 20,000 feet in six minutes. A sub-variant of the G-10 was the Bf109G-10/R2 which featured an armament of two MK 108 cannon in a ventral tray. To extend the range the cannons were deleted and a long-range tank fitted. The G-10/R2, together with the /R6, was modified by the addition of an increased area fin and rudder in an effort to counteract the vicious torque experienced on take off. The longer tailwheel oleo increased the ground space during take off, and an improved radio completed the modifications.

Bf109G-10/U4 with large auxiliary fuel tank.

The 100G Trainer

A second attempt at producing a two seat trainer, with pupil and instructor in tandem, resulted in the G-12. The student occupied the front cockpit. There was no production line established for this variant, they were built on an ad hoc basis, but local modifications at the maintenance units produced enough aircraft to satisfy demand. A fighter-bomber, the G-14 was a potent machine and could be powered by a variety of DB engines, ranging from the 605A, and including the AM, AS, ASB, ASD and ASM. A 'Galland' hood was standard, this being developed over a period of time. It was not as satisfactory as the clear, bubble hood that was adopted by American and British fighters, but did have less framing than the normal, sideways opening unit.

Armament of the G-14 was varied. One MG 151/20 cannon mounted over the engine block plus two MG 131, 13mm machine guns. Additional ordnance was a pair of MG 151/20 cannon, two WGr 21 rocket installations. One SC250 bomb could be carried on a ventral rack. A tropical variant of the G-14 was the G-14/Trop; the G-14/R2 had the increased area fin and rudder, while the G-16, hastily conceived for the ground attack role, carried additional armour for pilot, engine and fuel tank.

By the time the G-16 had appeared the basic 109 airframe was outdated, but there was not the time to introduce a new Messerschmitt piston-engined design. More performance was demanded by the *Luftwaffe* High Command and the decision was taken to utilise the basic 109 design and make use of a higher performance engine. To get the proposed variant, designated the Bf109K, into service as quickly as possible, a number of G-models were converted on the production line to the new standard, and the first true production models, the K-2 and K-4, now powered by the DB 605 ASCM/DCM engine producing 1,500hp, appeared in the closing months of 1944.

MW-50 injection was standard and this increased available power for combat or emergency speed, to 2,000hp. The main armament of both Marks was two MG 151 20mm cannon, plus a single MK 103 or 109 30mm weapon. The K-4 had a pressure cabin. A third version, the K-6, had the same DB engine but its armament varied in consisting of two MG 131 machine guns in the upper engine cowling; a 30mm Mk 103 firing through the propeller spinner, and could also be fitted with a pair of MK 103s carried under the wings.

Prototype Bf109G-12 two seat trainer.

Ground crew load camera in Bf109G PR variant.

The Allied advance through France had proved irresistible and German forces were driven back into the homeland. Despite massive damage and interruption to production facilities output of the 109 fighter actually increased. What was lacking was trained pilots and fuel. In early 1945, just a few months before the German surrender, the Messerschmitt organisation produced the final version of this versatile aeroplane. The Bf109K-14 was fast and powered by a DB 605L, plus MW-50 power boosting, and could reach a speed of 450mph. Two MG 131 13mm machine guns and a single MK 108 30mm cannon made up the offensive power.

But it was far too late as Allied Forces were now over running German territory from the East and West. Gruppe Stab of II./JG52 managed to collect a couple of the new fighters but there is little record of them in action. A variant designated the 109L was proposed, with the Junkers Jumo 213E engine. To accept this modification wing span was increased to allow the aeroplane to operate at a greater altitude, and the fuselage cross section enlarged. A maximum speed of 474mph was quoted. An experimental 109, the S, was to have featured the 'blown flap', super circulation, to improve low speed characteristics when the increased weight was a handicap. It was never completed.

The Messerschmitt company was immersed in the closing months of the war in the production of its 262 jet aeroplane, and in an effort to improve the 109 a proposal was put forward for the installation of the Jumo jet engines. It was decided to modify a normal 109G airframe but it would have resulted in a major re-design. The project, 109TL, was also abandoned.

Before leaving this short history of the G-series mention must be made of the twin 109 fuselage concept, a bizarre effort to produce a heavily armed bomber destroyer. A prototype was completed in 1943 but it was damaged in an air raid and never proceeded with.

Construction of the Bf109G-6

The wing, although of all metal construction with a single main spar built up of 'T' section flanges, was a flimsy affair despite a sheet web reinforcement at the intersection of the ribs. Main spar was located at some distance from the leading edge, and this was to allow the main chassis to retract into the wells. This feature, like that of the Spitfire, was a weakness that led to wing collapse when overstressed. Wing strengthening had to be employed in both the British and German fighter. Stressed-skin, flush riveted light alloy covering, together with fabric covered light alloy ailerons completed the wing structure. The wing halves were secured to the fuselage at three points, with the leading edge attachment consisting of a stout, steel forging which also housed the undercarriage leg and base of the engine mounting. A four section Handley Page-type trailing edge slotted flap was installed, and the leading edges were also slotted.

Fuselage was an all-metal, oval section, monocoque structure with top hat section transverse frames and 'Z' section stringers. A flush riveted stressed skin covered the inner construction and the production engineering was regarded as excellent. A BMW DB 605A-1, twelve cylinder 60 inverted Vee liquid-cooled in-line engine, constructed by the

Wing riding was popular, Bf109G-6.

Head on view of Bf109G shows huge size of air intake.

Daimler Benz A.G. Company, provided the necessary urge and delivered 1,475hp at take-off. This was translated into forward movement and lift through a V.D.M., electrically operated, constant speed, 3-bladed propeller. The normal complement of armament was two 13mm MG 131 machine guns with 300 rounds for each gun, and they were mounted in the upper engine cowling and synchronised to fire through the airscrew. A hard hitting, single 20mm MG151/20 cannon, with a 150 round magazine was mounted directly above the engine and fired through the spinner. A Field Conversion Set provided for two 20mm MC151/20 cannons, with a 120 round magazine for each gun, in underwing gondolas.

The tail section was a cantilever structure of mixed metal and plastic, with a stressed skin covering on the fixed surfaces and fabric covering the rudder and elevators. The undercarriage retracted outwards with the pintle attached to the wing/fuselage junction. The main wheels moved up sideways, and inwards, into the wings ahead of the main spar when in the fully retracted position. Like the Spitfire it required careful handling on unprepared fields and was prone to collapse if any large side movement was applied. The tailwheel was a semi-retractable unit.

CHAPTER EIGHT

Gustavs For The Luftwaffe

At the end of summer 1942 a total of almost a thousand Bf109s were in service with various units of the *Luftwaffe*. More than half of this total was the l09G. The first unit to equip with the Gustav was the 11th *Staffel* of JG2 Richthofen Geschwader. This unit was newly formed and together with 11./JG26 operated the new variant in France. II./JG2 had been established in late May 1942 from a nucleus of personnel from *1 Staffel*. Equipped with the Bf109G-1 it was to become a special, high altitude, interceptor unit.

Other units to receive the G were III./JG26, which also flew a small number of Fw190s, together with *Staffeln* in Russia, the Middle East, the Balkans and Norway. By the winter of 1942, the 109G replaced the F model, and it, together with the Fw190A, formed the equipment of the *Luftwaffe Tagjagdflieger* (day fighter force). In spring 1943, the Eighth Air Force of USAAF bombardment of the German homeland had reached alarming proportions as unescorted B-17 and B-24 bombers were used in massive, daylight attacks on military targets. Despite the desperate need for fighters on the Russian front III./JG54 was withdrawn on 27 March 1943 and a new *Geschwader*, JG11, was formed from personnel provided by I and III./JG1. The date was 1 April 1943, an inauspicious day on which to launch a new fighter unit.

The American bombers flew in close formation and their heavy .5 calibre Colt machine guns delivered a heavy weight of metal to the defending Messerschmitts and Focke Wulfs. New methods of defence had to be conceived, refined and developed, including the technique of attaching a 250kg bomb to the underside of a Bf109. After reaching altitude above the bombers the pilot would release his bomb, equipped with a sensitive time fuse, on the formation below him, endeavouring to kill as many of the four-engined attackers as possible.

The new system of attack was developed and tested by *Lt* Heinz Knöcke of 5./JG1, and he flew the first sortie against a B-17 formation that was attacking Wilhelmshaven on 22 March 1943. The result of the encounter was one bomb dropped, one B-17 destroyed. The idea left the *Luftwaffe* chiefs unimpressed and other methods were called for. These developed as heavy calibre cannon (30 and 40mm, similar to the Vickers 'S' and Rolls-Royce anti-tank weapon used on the Hurricane Mark IV). 210mm rocket tubes proved to be quite successful and were used operationally by *Staffeln* of JG1, JG3 and JG26.

Operation *Barbarossa,* the attack on Russia by its erstwhile ally, had started well for the German Forces. But, as Russian resistance stiffened, the Bf109G was a welcome sight over the many battlefields to the soldiers in the field. However, by the summer of 1943 the tide of victory had turned to the beginnings of defeat. In an effort to stem the Russian advance

Bf109G of II./JG 52, Russia. Barkhorn's mount.

Bf109G-6/Trop. Tropical filer and large fairing.

Bf109G of JG 51 in Russia. Yellow on upper wing tips.

the German High Command, under direct orders from Hitler, launched Operation *Zitadelle* in the Orel-Bilegorod area. It was a desperate move and proved to be the final, major German thrust in the East. The Russian armour was attacked from the air by Bf109Gs, Fw190s, Ju87s, in an attempt to make a breakthrough. Eight *Jagdgruppen* – II and III.JG3; I, II and IV.JG51; III.JG52; III./JG54, JG3 and 52 *Gruppen* were Bf109G units, and from the first day of the offensive massive victory claims were reported. On the first day of fighting no less than 432 Russian aircraft destroyed against a total German loss of 26. But replacements by the Russians appeared to be inexhaustible and the offensive finally stalled as the tired German troops and aviators could no longer sustain the pressure.

In other war theatres it was the same story as the tide finally turned in the Allies favour. Gustav-equipped *Gruppen* of *Luftflotte* 2 came under severe pressure in Southern Italy and Sicily. Stab, I and III./JG27, II/JG51, Stab, I, II and III./JG53 and Stab, I, II and III./JG77 could not resist the might of Allied Air Forces. In Northern Europe Norway and Finland JG5 had four *Gruppen* equipped with the latest model of the 109G, the G-2, and lost I and II *Gruppen* to the Russian front in January 1944 as the situation was desperate, but when the Second Front opened on D-Day, June 1944, they were moved rapidly to help cope with the new invasion. III./JG5 remained in the North to act as part of the rearguard covering the withdrawal of German forces from Finland. IV./JG5 was given the task of defending Norway and Denmark in company with IV./ZG 26, and a fighter training unit, 4G 102 equipped with Bf109 G-12s.

The Eighth and Ninth Air Forces of USAAF were stepping up the weight of their attacks against targets over the whole of Europe, and losses to units of the *Reichsverteidigung* to the P-51s, P-47s, Spitfires and other allied fighters rose to alarming numbers. The numbers of raids was increased when, on 1 August 1943, a large force of B-24 Liberators from the US 8th and 9th Air Forces left their bases in North Africa to mount an attack on the Rumanian oilfields at Ploesti. I./JG4 under *Hptm* Hans Hahn based at Mitzil, 20 miles east of Ploesti, equipped with the 109G and Fw190s, slashed into the American formation, meeting the incoming bombers in force, harassing the four-engined giants and keeping up the attacks as they headed for home. Fifty-four bombers never made their way back to base. It was a salutary lesson for the American elements and they were to overcome the defenders of Germany with the introduction of the long-range fighters equipped with overload fuel tanks.

Late in July 1943, five further *Jagdgruppen* were withdrawn from bases in Italy and Southern Russia to Germany to supplement the aircraft of JG1, JG2, JG11, JG26, III./JG54 and ZG 26 then engaged in the defence of the Fatherland. II./JG27 under *Hptm* Werner Schroer was transferred from Vibo-Valentia in Italy to Weisbaden-Erbenheim, II./JG51 under *Hptm* Karl Rommell from Sardinia to Neubiberg near Munich and the entire *Jagdgeschwader 3* under *Obstlt* Wolf-Dietrich Wilcke from Southern Russia.

On 17 August later that year the USAAF tried again and sent a large group of B-17s and B-24s – 376 bombers – against Schweinfurt and Regensburg. Among the many German fighter units waiting for the Americans were five *Jagdgruppen* transferred from Italy and Russia to join JG1, JG2, JG11, III./JG54, ZG 26. 60 of the attacking bombers were destroyed and many damaged. But, it did not deter the Americans for they launched a second attack, against Schweinfurt this time, on 14 October, losing a total of 60 out of 291 dispatched.

The American raids continued with an attack against Amsterdam on 3 November. This raid was different as it was escorted by protecting fighters and the German pilots were given a foretaste of future events. The *Luftwaffe* was to be put under extreme pressure when the long-range American and British fighters were roaming freely over Germany. Combat was almost suicidal as the Allied pilots were better trained, equipped with the latest variants of the Mustang, Thunderbolt, Spitfire, and new types such as the Tempest and Typhoon proved more than a match for anything but the Fw190D or the Messerschmitt 262 jet fighter. Lack of training was to result in a high mortality rate among the German pilots. Fuel was in short supply and training restricted.

The 'Gustav' in Foreign Service

Like all the major aircraft companies of the 1930s Messerschmitt was allowed to sell his new fighter to overseas customers. The fighter was to serve in numerous countries and the majority preferred it to its contemporary, the Supermarine Spitfire. The G was issued to Germany's allies, such as Bulgaria, who purchased 145 109Gs They were delivered to the Royal Bulgarian Air Force's Sixth Fighter Regiment at Karlovo and used in the defence of Sofia against the advancing Russians in April 1944.

Croatia, which became part of Yugoslavia when Tito became master of the many smaller states in that region, took delivery of a small number of Bf109G-10s. The Croation Air Force operated the type, and a small Croatian unit was set up inside *Jagdgeschwader 52* as its 15th *Staffel*.

Fifteen Bf109-Gs were despatched to Czechoslovakia in 1944, a move that would misfire when the country fell to Russia. The Czechs had no love for the German forces,

Bf109G.

but despite this plans were made for licensed production of the fighter. The Avia factory at Prague-Cakovice was selected to produce the 109G-14 but none were completed when German forces withdrew before the advancing Russians. After the war the Bf109G-14 production line at Cakovice was re-established and Messerschmitt's fighter appeared as the C-10. A two seat trainer was constructed, but there was no demand for this variant, called the Avia C-110 apart from those required for the Czech nation. The single and two seat variants entered service with the Czech Air Force as the S-99 and CS-99 respectively in small numbers, but supplies of the of German DB 605 engine were not available and production halted.

An alternative 1,350hp Jumo 211 F led to some re-design of the engine mounting and compartment. A broad, paddle blade propeller absorbed the power and the aeroplane went into production as the C-210 'Mezec'. It was a bad design but, surprisingly it was accepted for service with the Czech Air Force as the S-99, together with a two seat trainer version, the CS-199. The fledgling Israeli Air Force, eager to build its complement of fighters, purchased a number, but pilots were to complain of its characteristics. When Egypt attacked the new state, the Bf109 was to meet the Spitfire once again as the Egyptians were operating the Supermarine product. The S-199 and CS-199 remained operational with units of the Czech National Security Guard until 1957.

Finland, due to being so cruelly invaded by Russian only because it was a weak neighbour, was a staunch ally of Germany, and it must have been a bemused Air Force that turned from flying the Hawker Hurricane to using the 30 Bf109G-6s and 132 Bf109G-6s, which were diverted to Finland during 1943 and 1944 for use against the

Russians. The first unit to take delivery was the elite HLeLv 34 commanded by Major Eino Luukanen based at Utti. Up to the end of August 1944, the unit scored 270 victories for the loss of 11 pilots and 22 aircraft. By September 1944, all Finnish fighter units were equipped with the Bf109G, including a few G-14s which the retreating *Luftwaffe* donated to their ally.

Hungary was invaded by Germany and eventually fought with its forces. 59 German-built and approximately 700 home-produced Bf109Gs were delivered to the Royal Hungarian Air Force. The first units to be equipped were the 5/1 and 5/2 Squadrons of the 5/1 Fighter Group. 5/2 was to be re-designated as the 102nd Independent Fighter Squadron, and in June 1944 this unit expanded into a Group of two squadrons. A new Bf109G unit, the 101st Fighter Group was also formed to defend Hungary against Russian forces under the command of *Obstlt* Aladar Heppes. It was later expanded into a Regiment of six squadrons. From the spring of 1945, with the war almost at an end, Hungarian fighter squadrons concentrated on ground attack duties against the Russian Army with little effect after diverting from bomber interception duties. The end came when lack of fuel grounded the remaining Messerschmitts.

Seventy Gustavs (mainly G-8s) were exported to Romania, but this air arm was so weakened by early 1944 that it was unable to put up any effective opposition against Allied forces. Sixteen Bf109Gs were also built by the I.A.R. factory at Brasnov before it was destroyed in an Allied Air attack.

The Spanish Air Force received its first deliveries of the Bf109 and the Bf109B in March 1937. Forty-five Bf109Bs (Spanish designation C-4), fifteen Bf109Es (C-5),

A Finnish Air Force Bf109. Note underwing markings. Yellow nose and white fuselage band.

Bf109G prepares for take off.

ten Bf109Fs (C10) and twenty-five Bf109G (C12) airframes were delivered to Spain during the Second World War. After the war the Bf109G airframes were equipped with the 1,300hp Hispano Suiza 12-Z-89 engine, and carried the designation of HA 1109-JIL. The first production aircraft made its initial flight on 2 March 1945. The first genuine Spanish production version was the HA 1109-KIL which flew in March 1951, and it differed from the first machine in being powered by the Hispano Suiza 12-Z-17 engine. The HA 1110-Kil was yet another two-seat conversion and flew for the first time in October 1953. The HA 1112-KIL was equipped with a combination of rockets and cannon.

That constant neutral, Switzerland, accepted twelve Bf109G-6s and they were delivered to Fliegerkompagnie 7 coded as J-701 to J-712 inclusive. The Swiss Air Force added to its mixed bag of impressed aeroplanes when two more Gs strayed over Swiss territory. They were to become J-713 and J-714. The aircraft were rarely serviceable as German replacement parts were virtually unobtainable.

Bf109G-2
Leading particulars

Wing span: 32ft 6.5in, area 174.376sqft. *Length:* 20ft 0in. *Height:* 8ft 2.5in. *Weights:* tare 4,968lb, gross 6,834lb, max permissible 7,055lb. *Max speed:* 317mph @ S/L, 365mph @ 9,840ft, 369mph @ 16,400ft, 406mph with GM 1 boost @ 28,540ft. *Ceiling:* 39,370ft. *Rate of climb:* 4,590ft/min @ S/L. *Height to time:* 19,685ft in 3.2mins, 32,810 in 12mins. *Range:* (internal fuel) 340 miles, (with drop tank) 528 miles. *Engine:* Daimler Benz DB 605A-1 of 1,475hp @ T/O, 1,355hp @ 18,700ft, 1,250hp with GM 1 @ 27,980ft. *Armament:* One 20mm MG 151 cannon and 150 rounds, plus two 7.9mm MG 17 m/gns and 500rpg.

Bf109G-6
Leading particulars

Wing span: 32ft 6.50in, area 174.375 sqft. *Length:* 29ft 8in. *Height:* 8ft 6in. *Weights:* tare 5,900lb, gross 6,950lb, max permissible overload 7,500lb. *Max speed:* 387mph @ 22,970ft, 338mph @ S/L. *Rate of climb:* 19,000ft in 6 mins. *Service ceiling:* 38,500ft. *Absolute:* 39,750ft. *Range:* 450 miles @ 330mph @ 19,000ft, 615 miles @ 260mph @ 19,000ft. *Engine:* One 1,475hp Daimler Benz DB 605 A-1, twelve-cylinder, liquid-cooled in-line unit. *Armament:* Two 13mm MG 131 machine guns mounted in upper engine cowling, 300rpg, one 20mm MG 151/20 cannon mounted above engine and firing through spinner, 150rpg plus two under wing mounted MG 151/20 cannon of 20mm, 120rpg.

Bf109G

Bf109G-6, North Africa.

Bf109G with drop tank.

Bf109G-2, II./JG 54 'Grunherz', Tunisia, 1942-43.

Bf109G-2, II./JG 54 'Grunherz', Russia 1942-43.

Air gunner's view of Bf109G as it approaches for attack.

The photographer lines up his view-finder to film these desert Bf109s.

Belted ammunition is fed into magazine. II./JB 54 'Grunherz'.

Ground staff carry out routine maintenance duties on these Bf109Gs.

Note the green heart fuselage marking and belly fuel tank on this Messerschmitt.

This Bf109E is painted white overall for the Russian winter.

Ground crew direct the pilot as he ploughs through the mud at the time of the Allied landings in June 1945.

CHAPTER NINE

Swansong of the 109

As the numbers of raids, and weight of bombs dropped, by the USAAF increased during the daylight hours, and the RAF, with its high flying Mosquito Pathfinders, kept up the pressure during the nocturnal period, so did the urgent need for single seat fighters by the *Luftwaffe* increase. The Germans possessed za number of twin-engined types, but they were in short supply, and production did not increase fast enough to satisfy demand. The Allied threat from the air had to be met and, hopefully, brought to an end, and a high altitude fighter, produced in quantity, could help remedy this situation.

A crash programme was agreed between the RLM and Messerschmitt, for the development and production of a high altitude variant of the 109, as it was in series production and the company considered it could produce a suitable version and get it into production in approximately twelve months from initiation.

In early 1943 an interim design was agreed upon and the designation Bf109H allotted. Delivery of production models to be ready by the agreed twelve month date. Messerschmitt took a standard 109F and inserted a parallel chord centre section. This modification increased wing span to 39 feet and raised the operating ceiling. The RLM's original demand had been for a ceiling of 40,000 feet, but this was soon to be raised by an additional 9,000plus. To lift the projected fighter to that height a new model of the DB 603 was planned. Known as the 628 it featured a turbo supercharger similar to that of the American B-17.

Messerschmitt, not wanting a re-worked 109, submitted a proposal for a totally new design, the Me209 II, but RLM wanted its high altitude fighter in a great hurry and was not prepared to wait for a new aeroplane. Such an aeroplane would take many months to design and construct and a prototype would not be available until much further into the future. The company was instructed to proceed with the 109H, but to embody a number of the features of the 209, such as the engine, and to have a prototype flying at the earliest possible moment.

A DB 628 engine was flown in a 109G in May 1943 and approved. However, the new engine had certain disadvantages – it was bulkier, heavier and much additional design was necessary if the -H mounting and cowling could accommodate it. The design office extended the forward fuselage by 2.5 feet and only by moving the entire wing forward by 10.5 inches could the centre of gravity be re-established. Enlarged tail surfaces were required to ensure stability. The extended centre section also moved the undercarriage oleos outwards, and twin radiators were installed in the centre section. A second test 109 was modified to test the installation and, as the 109H, it flew in June.

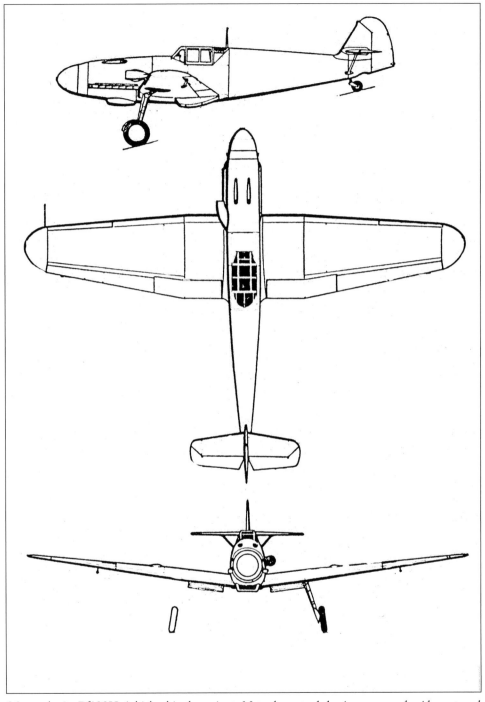

Messerschmitt Bf109H-1 high altitude variant. Note the extended wing span and wide, outward retracting undercarriage.

In the meantime work was proceeding on a number of 109F airframes. Pre-production 109H-1s were ready for flight testing in the winter of 1943/44, such was the urgency for them. With the new engine a ceiling of 47,500 feet was attained with speeds of 455mph possible in a dive. Problems were to arise, however, such as a wing flutter and stability, and, while the test aircraft were being modified, small numbers of the Focke Wulf TA152H became available.

A Junkers Jumo variant of the Bf109H was proposed as were other versions, the Bf109H-2, H-3 and H-4, all differing in armament and duties. All activity on the Bf109H ceased when the true prototype airframe was destroyed in an Allied bombing raid.

Bf109H (V55)

Leading particulars

Wing span: 43ft 6in, area 235.729sqft. *Length:* 33ft 7.5in. *Height:* 10ft 7.5in. *Weights:* tare 6,338lb, gross 7,804lb. *Max speed:* 427mph @ full boost height, 367mph @ 19,685ft. *Ceiling:* 44,290ft, cruise 292mph. *Engine:* Daimler Benz DB 605B of 1,600hp @ T/O. *Armament:* one 30mm Mk 108 cannon with 60 rounds hub mounted and two 20mm MG 151 wing mounted cannon and 200rpg.

The Bf109K Series

The first of the pre-production models of the new 109K-0 appeared in late September 1944, and was to differ from the later models of the G-series in having a raised engine cowling and larger spinner. It retained the Galland hood and an increased area empennage. A 30mm cannon was standard armament, plus twin 13mm MG 151s of 15mm. Series production models began leaving the factories as the 109K-2 and K-4, with the -4 having a pressure cabin.

The K-6 followed and was destined as a bomber destroyer with an additional pair of Mk 103 cannon in under wing gondolas, plus a pair of 13mm machine guns. Final deliveries from Messerschmitt were small batches of the K-14, just fourteen days before the German surrender. Its DB 604L engine provided 1,725hp at low altitude and 1,350hp @ 31,000 feet. Maximum airspeed was 452mph @ 19,600 feet. Armament was meagre by contemporary standards, just a single 30mm cannon supplemented by twin 13mm machine guns.

Late model Bf109K takes off from a Russian airfield.

Captured late model Bf109K with RAF markings. This aircraft also has a 'Galland' cockpit fitted.

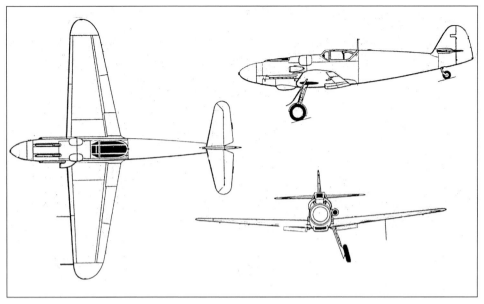

Bf109K-2 had a pressure cabin for high altitude operations.

Bf109K-4

Leading particulars

Wing span: 32ft 8.5in, area 173.299sqft. *Length:* 29ft 0in. *Height:* 8ft 2in. *Weights:* tare 6,834lb, gross 7,475lb. *Max speed:* 378mph @ S/L, 452mph @ 10,685ft, 435mph @ 24,610ft. *Ceiling:* 41,000ft. *Rate of climb:* 4,880ft/min @ S/L. *Height to time:* 16,400ft in 3mins, 39,370ft in 10.2mins. *Range:* 356 miles @ 19,685ft. *Engine:* Daimler Benz DB 605ASCM of 2,000hp @ T/O, 2,030hp @ 1,640ft, 1,800hp @ 16,400ft. *Armament:* one 30mm MK 103 or 108 with 60rpg and two 15mm MG 151 cannon with 220rpg.

The Bf109Z 'Zwilling'

Following the precedent set by Heinkel with his successful twin fuselage He111H bomber for use a target tug, Messerschmitt undertook to adapt the Bf109 airframe, not for glider towing but as the basis for a heavy *Zersterör*. RLM approval was forthcoming and two 109F airframes allocated for prototype conversion.

The design was two standard 'F' fuselages and outer wing section, plus a newly designed centre section. To cool the two engines the four main air radiator intakes were installed, two in the new centre section and two in the wings.

Production planning was to use the 'G' airframe and either DB605A or Jumo213E engines. Armament was light considering the aircraft's role with two hub mounted 30mm MK 108 cannon and a further two in under wing gondolas. A single 1,102lb bomb was to be carried on a rack under the centre section and one 551lb bomb under each fuselage.

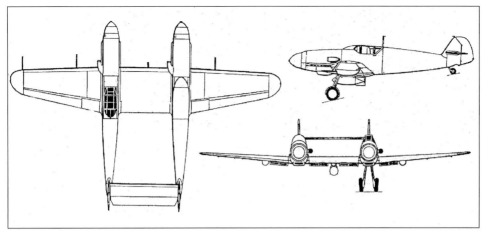

Like the North American Mustang the Bf109K-2 was an attempt to prolong the life of the fighter simply by joining together two fuselages, a new centre section wing and original outer panels. It was also used as a glider tug.

The undercarriage had to be a compromise with two oleos retracting into the centre section and two in the outer panels.

A single pilot was all that was required and the starboard cockpit was faired over. The only prototype was finished in 1943 but it, apparently, was damaged in one of the many air raids. The programme was abandoned the following year. Germany did not require aircraft. What it needed was the fuel to fly them.

Bf109Z

Leading particulars (Estimated)

Wing span: 43ft 6in, area 249,70sqft. *Length:* 29ft 2in. *Height:* 8ft 10in. *Weights:* tare 16,050lb, gross 17,882lb. *Max speed:* 374mph @ S/L, 462mph @ 26,250ft, cruise 428mph economic. *Rate of climb:* 5,080ft/min @ S/L. *Range:* 1,050 miles, max 1,240miles. *Engines:* Junkers Jumo 213E of 1,750hp @ T/O, 1,320hp @ 29,530ft. *Armament:* four 30mm MK 108 cannon and one 30mm MK 103. Bomb load of 1,102lb and two 551 missiles.

Before leaving this history mention has to be made of a number of other 109 projects such as the Bf109E-Reihi of 1939 with a DB601A engine, two 20mm MG/FF cannon in the wings and two 7.9mm MG 17 m/gns in nose section. The prototypes Bf109V14 and V15 were to be used for this exercise. Abandoned.

The Bf109-L-Reihe was the prototype for the Me209-II with the Junkers Jumo213E engine and GM-1 injection. There was also a projected Bf109-S-Reihe about which little is known.

Second Stage Variant, the Me209

The Me209 fighter was originally developed as an Air Speed Record entry and this was a development of the early Bf109 fighter. Determined to reveal to the aviation fraternity that Messerschmitt's new fighter was better than any contemporary type three highly modified prototypes were built with the aim of producing a record breaking aircraft designed around the in-line evaporatively cooled Daimler Benz 601 engine. Lessons learned from the venture would be incorporated into the forthcoming new Messerschmitt fighter.

It was designated as the Me209 and when completed could be seen as having little connection with the Bf109 fighter. With a wing span of just 25ft 7in and a gross weight of 5,545lb, wing loading was a high 48.746lb/sqft. The Me209 VI was ready for trials during June 1938 and it flew for the first time on 1 August 1938. The test pilot reported unfavourably on performance and he detailed no less than seventeen major faults.

The second prototype V2 flew on 4 April 1939 and came to grief when the oil cooling system failed and the engine stalled. The pilot managed to glide before crashing and writing off the aircraft. A new engine, the DB 601 V10 of 1,559hp, using special fuel, was available and installed in the prototype VI. Maximum power could be raised to 2,300hp when methyl alcohol was injected. The VI (called the Me109R for propaganda purposes) flew with the new engine on 26 April 1939 when the first attempt by the aircraft on the World Air Speed Record was made. A maximum speed of 469.22mph was attained, just 5.3mph than the fastest ever recorded by rival machines.

The third 209 prototype was the V3 which was completed and flew in May 1939 and made its first flight after the fourth prototype V4 had flown. The latter had provision for an armament of two 7.9mm MG 17 m/gns and a hub firing 20mm MG FF/GM cannon. The fin/rudder assembly was increased in area to provide lateral stability. A shallow, low drag radiator system replaced the evaporative. It had a wing span of 33ft 0in and a loaded weight of 4,806lb.

The following year the V4 was flying with a SDB 601N of 1,200hp @ T/O and emergency of 1,270hp @ 16,400ft. Armament was specified as two 7.9mm m/gns in the engine cowling and a hub firing 30mm MK 108 cannon, plus a further two in the wings. Weight rose dramatically and performance suffered. To assist cooling a deep radiator cowling was installed and maximum speed was reduced.

The RLM and Messerschmitt agreed that the aircraft was now unsuitable for any record attempt but it was suggested that the aircraft could be modified and used as a development of the Bf109G. As such the prototype was now referred to as the Me209 II using the Bf109G wing and components of the 209 design. A Daimler Benz 205 engine was made available and was cooled via an annular radiator. A first prototype was designated as the Me209 V5.

Wind tunnel trials revealed that the wing was not suitable for the high speeds required of the new fighter and a different aerofoil was adopted. Armament was specified as two 13mm MG 131 m/gns in the nose section but they were eventually transferred to the wing roots. The V5 flew for the first time on 3 November 1943 with a DB 603A engine, but this was replaced by a 603G. It also had a larger fin and rudder.

The Daimler Benz in-line engine was eventually discarded in favour of the Junkers Jumo 213E, also an in-line, which altered the configuration. However, the changes were

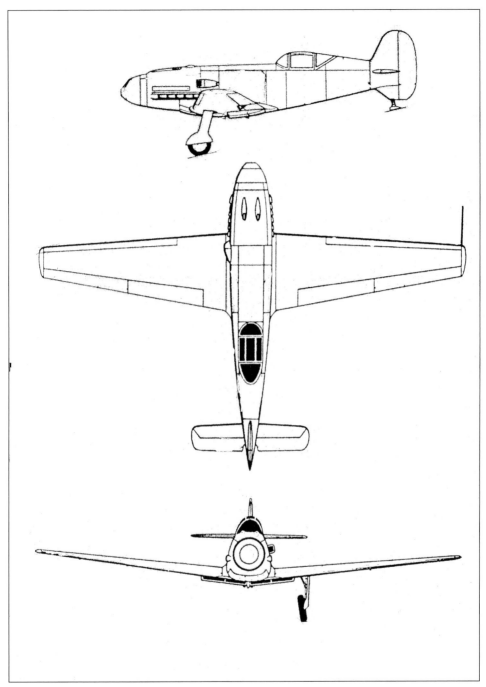

The British and German Air Ministries adapted the Bf109 and Spitfire respectively as record breakers. It was the Me209V4 that captured the record as the Spitfire was never entered for the contest.

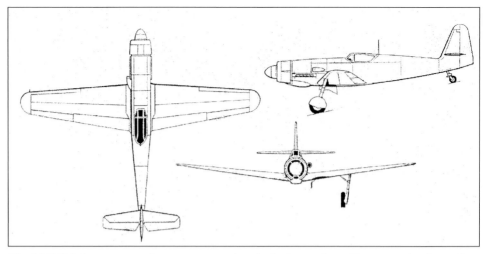

The Me209A-1 was yet another attempt to update the basic aeroplane. Gone were the elegant lines produced with the Daimler Benz in-line engine as it was replaced by the Junkers Jumo radial.

completed by the end of November 1943. Another prototype, the Me209 V6 flew in April 1944 with a Jumo 213E of 1,750hp at T/O. To boost power, at 38,000ft GM 1 nitrous-oxide was injected.

Production was slow to build up, the first being the Me209A-1 with Daimler Benz 603G and an armament of a single hub firing 300mm MK 108 cannon or two 20mm MG 151s in the wing roots. The Me209A-2 had the Junkers 213E engine and was armed with two 20mm MK 151 cannon plus two 13mm MG 131 m/gns in wing roots. The Me209A-1U1 would have had the same armament as the 209A-2 plus two 30mm MK 108 cannons in the wings. The 209A-1U3 was proposed as a fighter bomber carrying either a 551lb or five 110lb bombs on fuselage racks. The Me209A-2/U3 and 4 used GM 1 equipment.

By the time this interesting series of fighters could be produced in any serious quantity Germany had been overrun and the war was over.

Me209 V5/A-1

Leading particulars

Wing span: 35ft 11in, area 184.6sqft. *Length:* 31ft 11in. *Height:* 13ft 1.5in. *Weights:* tare 7,360lb, gross 9,006lb. *Max speed:* 417mph @ 19,850ft, 463mph with GM 1 boost @ 39,300ft, cruise 323mph. *Ceiling:* 39,370ft. *Engine:* Daimler Benz DB 603G of 1,750hp, or Junkers Jumo 213E of 1,560hp. *Armament:* as above.

Me209H

It was during April 1943 that the first proposal for a high altitude fighter based upon the Me209 was proposed. To cope with the higher altitudes suggested the wing span would be increased to 43ft 6in. Engine specified was a Daimler Benz 628A or DB 603U with TKL 15 turbo-supercharger. Design was completed in the following October and work commenced on construction of the prototype 209H V1.

Due to delays it was ready in June 1944 by which time the Allied forces had landed in France and all work was abandoned in favour of production of the standard 209G.

Me209H V1

Leading particulars

Wing span: 43ft 6in, area 235.729sqft. *Length:* 32ft 5.5in. *Weights:* tare 6,636lb, gross 9,480lb. *Max speed:* 391mph @ 19,680ft, 460mph with boost at altitude, cruise 298mph. *Ceiling:* 44,290ft. *Engine:* Daimler Benz 627B of 2,000hp @ T/O and 1,325hp @ 34,200ft. *Armament:* one hub firing 13mm MG 131 cannon and four MG 151 20mm.

The Final Attempt – the Me309

Following the Battle of Britian the Messerschmitt design team was at full stretch with various lines of development of the Bf109 fighter. Many configurations were examined, and one aspect was to provide a more durable undercarriage to replace the rather flimsy outward retracting unit on the 109. A number of prototypes used for this programme had the wide undercarriage configuration, plus others with the same wide track units and a nosewheel.

Prototypes V23/V30 were used and were to be utilised in a new design and prototype called the Me309 V1, completed in June 1942. The new nosewheel arrangement raised many problems such as wheel wobble of the front unit. The first accident occurred in September 1942 during landing with a reverse pitch propeller. The second prototype, Me309 V2 flew on 29 November 1942 and this too was damaged during landing when the nosewheel collapsed.

The Me309 V4 flew with the DB 605B engine and carried a full armament of two 13mm MG 131 in the fuselage, a 20mm MG 151 and a MG 131 in the wing roots. Finally, a 30mm MK 108 cannon in the outer wing panels. Production started with the 309A-1 and 309A-2. The 309B-1 was a dive-bomber with cannon armament and two 551lb bombs on the wing fuselage junction. The Jumo 213H of 2,650hp @ T/O was an alternative engine but never installed.

The final design proposal was the Me609 consisting of two 309 fuselages with standard outer wing panels and a new centre wing section. It also had two fin/rudder assemblies and a connecting tailplane.

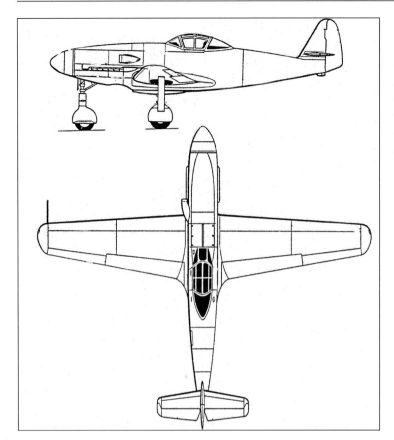

The final attempt to redesign the Bf109 was the Me309. It had a more powerful engine and what was more significant was the bi-cycle style of undercarriage.

Me309 V4

Leading particulars

Wing span: 36ft 1in, area 177,389sqft. *Length:* 32ft 7in. *Height:* 11ft 3in. *Weight:* gross 10,736lb. *Max speed:* 360mph @ 7,220ft. *Ceiling:* 37,400. *Time to height:* 5.2mins to 13,100ft. *Range:* 683 miles. *Engine:* Daimler Benz 605B of 1,475hp @ T/O. *Armament:* two 13mm MG 131 m/gns in fuselage, two 13mm 131s in wing, two 20mm MG 151 in wing and two 30mm MG 108 in outer wing panels.

CHAPTER TEN

Camouflage and Markings

The first Bf109B-1s entered service with II./JG132 'Richthofen' in early 1937 and with 2./J 88 in Spain in April of that year. The colour scheme of the fighter was light grey upper surfaces and pale blue grey unders. The national insignia comprised the *Balkenkreuz,* a narrow black cross outlined with white, open at the ends with a thin black outer trim line. This scheme appeared on the upper and under surfaces of the wings with the centre of the cross slightly inboard of the aileron and on the fuselage the vertical bar just aft of frame four. On the fin and rudder appeared the National Socialist insignia a black *Hakenkreuz,* or swastika, tilted on end and positioned centrally on a white disc on the thick red band.

The scheme remained standard until early 1938 when all upper surfaces were repainted in black/green *(schwarzgrun 70),* the only exception being the Bf109s of J.88 in Spain which retained the grey scheme but with the Condor Legion black cross on a white background on the fin/rudder assembly. With the introduction of the black/green scheme the red band and white disc were removed from the fin/rudder, the swastika remaining centrally positioned and thinly outlined in white.

The black/green scheme provided little protection against enemy strafing aircraft as a ground camouflage when dispersed on normal grass airfields, and with the war imminent over Europe in the late summer of 1938 an attempt was made to lighten the stark, near black appearance of the Bf109 with a disruptive splinter scheme pattern on all upper surfaces in dark green *(dunkelgrun 71).* Contemporary photographic evidence indicates that in the initial application of the splinter scheme a number of squadrons left the fuselage spine in black/green.

It can also be stated that the early splinter pattern varied to a considerable extent between separate aircraft on the fuselage sides, but to a much lesser degree on the wings and tailplane. During the autumn of 1938 the RLM must have issued a standard splinter scheme to both squadrons and factories and the result was that a constant pattern appeared. However, a large number of squadrons retained the black/green scheme until after the Polish Campaign in late 1939.

By the end of 1939 the splinter scheme was in general use with all operational units but the black/green scheme still lingered in the training squadrons. In early 1940 the defensive camouflage of the Bf109 underwent a further change, this time to the offensive. The scheme was employed during the Battle of France and most of the Battle of Britain and consisted of painting the fuselage sides and vertical tail surfaces a pale blue grey, with the upper fuselage decking left in the black/green splinter scheme to depths varying from the lower cockpit frames to a line parallel with the top of the fuselage cross.

During this change of scheme period the national insignia was also altered when the vertical bar on the upper wing crosses were moved just inboard of the aileron, the vertical

This Bf109E displays the symmetrical upper-wing splinter camouflage during the Battle of Britain in 1940.

Early model Bf109E in pre-war or early war month markings.

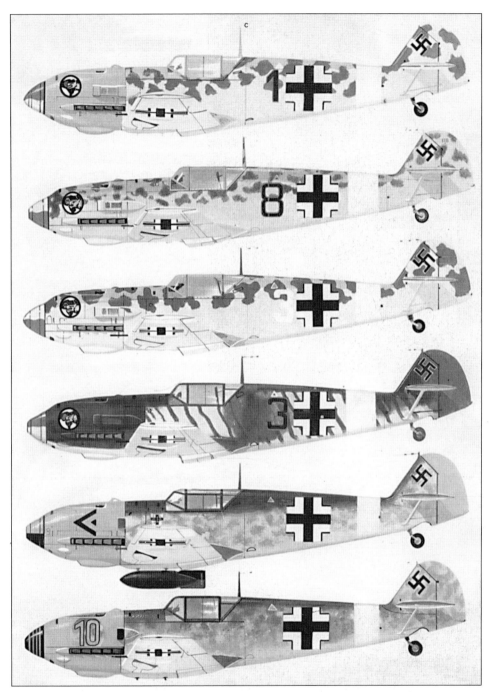

Top to bottom: *Bf109E-7/Trop, I./JG27, Cyrenaica, Libya, June 1941. Bf109E-4(N)/Trop, I./JG27, Derna, Cyrenaica, Libya, 1941. Bf109E-4(N)/Trop, I./JG27, Derna, Cyrenaica, Libya 1941. Bf109E-4(N)/Trop, I//JG27, Tmimi, Cyrenaica, Libya 1941. Bf109E-7,* Gruppen Kommandeur *III./JG27, Balkans, 1942. Bf109E-7, III./JG27, Balkans, 1942.*

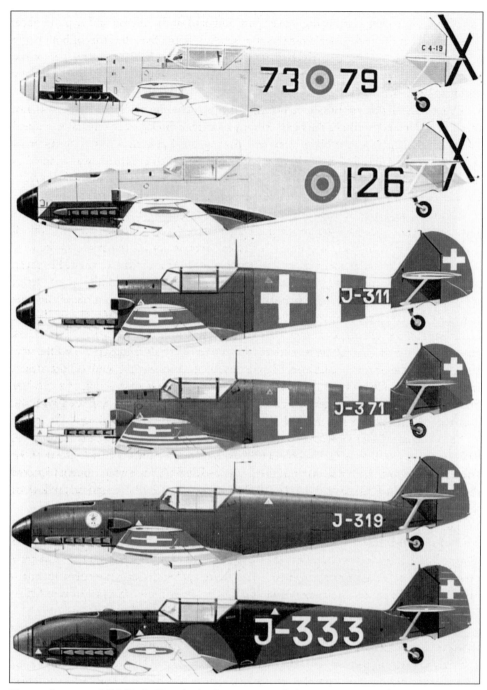

Top to bottom: *Bf109B-2, Escuela de Caza, Moron de la Frontera, Spanish Air Force, 1945. Bf109E-1, 25 Grupo, 23 Regimento de Caza, Spanish Air Force. Bf109E-1, Swiss Air Force. Bf109E-3, Swiss Air Force. Bf109E-3, Swiss Air Force. Bf109E-3, Swiss Air Force.*

bar of the lower wing cross being moved just outboard of the aileron and flap interface. The vertical bar of the fuselage cross was moved aft of frame five, the arms of both being outlined in a thick white area which had a thin black outer trim line. At this period the swastika was moved to the fin retaining the thin white outline which was normally outlines with a thin black line.

During the Battle of Britain it became obvious that the revised scheme was not practical and in later summer the high visibility blue-grey fuselage side scheme was toned down with, initially, a new black/green and dark green dapple in a variety of different applications. This was followed with dapple RLM grey which eventually was adopted as the standard scheme of the Bf109 for what remained of the war with exceptions such as the desert scheme and Russian variation to reflect the climate conditions they operated in.

However, the basic European scheme underwent several changes and additions at squadron level. The black/green scheme supplemented the RLM dapple grey on all European fronts, with many alterations in shades of colour for the remainder of the war.

In the Mediterranean area the first Bf109s arrived in Africa in the standard European scheme and eventually had to be radically altered to suit the conditions of desert operations. The scheme adopted was sand (a pink-fawn) on all upper surfaces and pale blue-grey unders. This scheme was to be altered as the squadrons decided what was the better camouflage for the areas they flew over. Some squadrons applied a light or dark green dapple on all the sand upper surfaces and a mixture of all sand uppers and the new change, and the various squadron adaptations flew. Towards the end of the desert campaign, as the Allies drove the Afrika Korps into Tunisia, the destruction of the Bf109 fighter rose to alarming proportions, and replacements from Europe appeared with the dark green/black splinter scheme as there was not time to change the colours.

Squadron code letters were seldom applied to aircraft of the *Jagdgeschwader* and, in common with other fighters such as the Fw190, the Bf109 displayed coloured numerals ahead of the fuselage *Balkankreuz*, the colour denoting the *Gruppe* and *Staffel*. Thus, a white numeral denoted the first *Staffel* in each *Gruppe*, i.e. 1, 4, 7 or 10 *Staffel*, black or red for the second, and yellow for the third.

The *Gruppe* itself was further denoted by the presence, or absence, of a further symbol after of the *Balkankreuz*. For example, *I Gruppe* aircraft displayed no such symbol; *II Gruppe* aircraft with a horizontal bar; *III Gruppe* a wavy or vertical line, and *IV Gruppe* a vertical cross or a disc. Further symbol codes were applied to staff members aircraft – *Staffelkapitän, Gruppenkommandeur, Geschwader Ia* (the Operations officer), *Technical Officer* and *Adjutant* codes adopted by all the *Geschwader Kommodore* varied unit by unit.

Other sections of the Bf109 were frequently painted in the *Staffel* colours, although this practice appeared and disappeared throughout the final years of the war. The under cowling, spinner, in part or overall, wing tips and base of rudder, were often painted in such colours, although in a number of instances the reason for the choice of a particular colour was often remote.

Excepting those *Jadgeschwader* which remained deployed in the West for long periods, it was common practice to redeploy German Air Force units from front to front according to strategic priorities, and this constant movement of aircraft undoubtedly accounted for the numerous – often obscure and possibly confusing – colour schemes and insignia appearing on German fighters. That they conformed to an often short-lived practice is undoubted but, viewed in retrospect, such schemes appeared highly confusing.

A note about 'radio codes' should be added as these were frequently applied on German fighters and might otherwise be confused with the *Geschwader* codes which, in any event seldom, if ever, appeared on fighters. These code letters were invariably painted forward and aft of the *Balkankreuz* and under the wings. They were normally applied for identification during flight trials, in particular prototypes by the manufacturers or delivery flights to operational units. Occasionally they were applied during development flying at Reclin and other Service trials establishments. Normally they were painted in black but, occasionally, in particular in the Mediterranean theatre, in white. No pattern of such codes has been listed, and it is believed that the only criterion that existed was the pairs of letters should not coincide with those adopted as established *Geschwader* codes.

A final word concerning the Russian campaigns. When Hitler's armies and air force crossed the frontier of Russia the fighters and all aircraft were painted in the familiar splintered dark green/black uppers and pale blue grey unders, plus the normal national insignia and *Gruppe/Geschwader* markings. As the campaign dragged into winter there was little time to apply a suitable scheme to reflect the conditions. As a result the aircraft were roughly painted with white on the upper surfaces in washable distemper and pale blue grey unders, and this deteriorated quickly in the harsh conditions in which they operated.

Some Bf109s were seen with other identifying marks, such as red, on engine cowlings and wing tips. But the situation was always haphazard – when the better weather approached the white was washed off and the splinter scheme not repainted.

Pilots of JG53 'Pik As' enjoing the grape harvest. Probably Italy 1945.

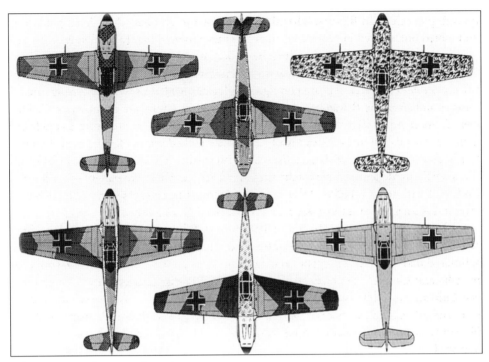

Bf109 upper surface camouflage schemes.

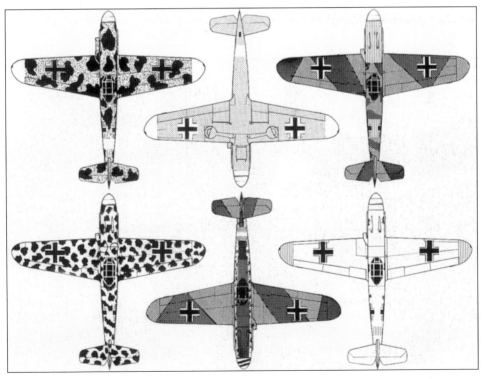

Bf109 camouflage schemes.

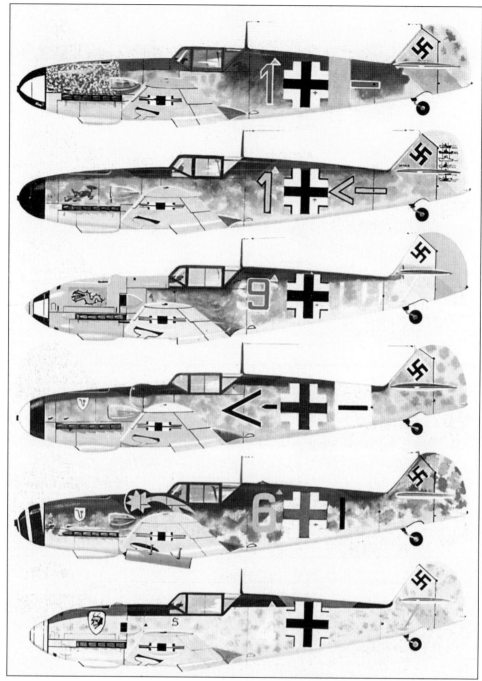

Top to bottom: *Bf109G-4 II./JG3* 'Udet', *France 1941. Bf109G-4, 10(Jabo).JG2* 'Richthofen', *France 1942. Bf109E-4, I./JG3* 'Udet', *France 1940. Bf109G-2, JG3* 'Udet', *flown by* Geschwader 1A, *Mediterranean, 1943. Bf109G-6, III./JG3* 'Udet'. *Bf109E-1, 2./JG26* 'Schlageter', *France 1940.*

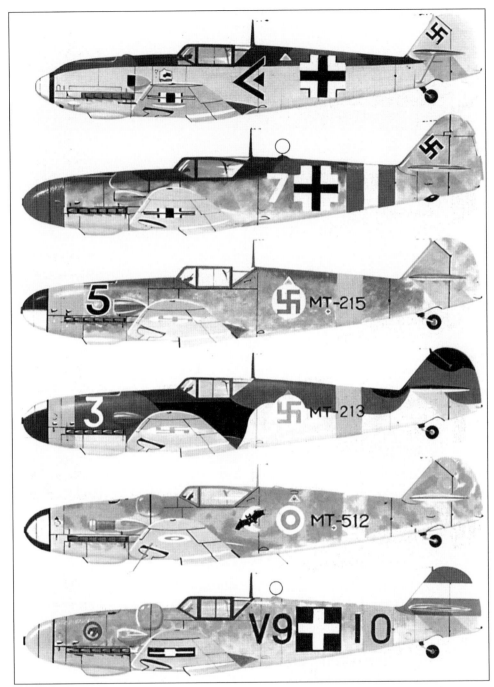

Top to bottom: *Bf109E-4, II./JG77, flown by the* Gruppen-Kommandeur, *1940. Bf109G-K-2, probably JG300, Home Defence, Germany 1945. Bf109G-2, 1/LeLv34, Finnish Air Force, 1943. Bf109G-2, 1/HleLv24, Finnish Air Force, summer 1944. Bf109G-10, night fighter, Flight HleLv31, Finnish Air force, 1948. Bf109G-6, 5/I Fighter Group, Hungarian Air Force, flown by Major Aladar Heppes von Bebayes.*

APPENDIX I

Messerschmitt Bf109 variants

The Supermarine Spitfire was produced in twenty-four versions and six Seafire variants, a total of thirty. There were many more 109s, but not all were produced in such large numbers as the F and G models. They have been grouped in this section of the history for convenience.

Bf109A (V1). First prototype with Rolls-Royce Kestrel engine.

Bf109 V2. Second prototype.

Bf109 V3. Third prototype.

Bf109 V4. With Jumo210A engines.

Bf109B. Pre-production 109B-0 with Jumo210D engine.

Bf109B-1. With Jumo210D.

Bf109B-2. With Jumo210E and 210G engine/s.

Bf109 V10. Prototype D-ISLU.

Bf109C. Developed from Bf109 V8.

Bf109C-0. Armament of four 7.9mm guns.

Bf109C-1 As above.

Bf109C-2. Armament of five 7.9 guns.

Bf109 V13. Prototype D-IPKY.

Bf109 V13. Special a/c powered by boosted DB601. Air Speed Record Holder.

Bf109D. Development of 109 V10/13 prototypes.

Bf109D-0. DB600Aa and armament of one 20mm cannon and two 7.9mm m/gns.

Bf109D-1. As above.

Bf109D-2. Armament of two wing mounted 7.9 m/gns.

Bf109D-3. Two 20mm MG FF cannon in wings.

Bf109 V14. (D-IRTT) prototype with injection DB 601A. Two 20mm and two 7.9mm guns.

Bf109 V15. (D-IPHR) as above but with one 20mm cannon.

Bf109E. With four 7.9 m/gns.

Bf109E-0. As above.

Bf109E-1. Fighter bomber.

Bf109E-1/B. As above.

Bf109E-2. Two 20mm and two 7.9mm guns.

Bf109E-3. One engine mounted 20mm cannon and four wing 7.9 mm m/gns.

Bf109E-4. As above but no cannon.

Bf109E4/B. As E-3 but no cannon.

Bf109E4/Trop. As E-3 but no cannon.

Bf109E-4/N. DB 601N engine.

Bf109E-5. Recco/fighter with two 7.9mm m/gns.

Bf109E-6. Recco/fighter as above.

Bf109E-7. As 109E-4/N plus provision for ventral fuel tank.

Bf109E-7/U2. Ground attack.

Bf109E-7/Z. GM-1 boosted engine.

Bf109E8. DB 601E engine.

Bf109E9. Recco/fighter.

Bf109F. As basic E with DB 601N engine

Bf109F-0. As above.

Bf109F-1. Armament of one 20mm and two 7.9mm guns.

Bf109F-2. Armament of one 15mm and two 7.9mm guns.

Bf109F-2/Z. With GM-1 engine boost.

Bf109F-3. DB 601E engine.

Bf109E-4. Armament of one 20mm and two 7.9 guns.

Bf109E-5. Recco/fighter with two 7.9mm guns.

Bf109F-5. BMW 801 test bed.

Bf109F-5. Jumo 213 test bed.

Bf109F. With butterfly tail unit. Also, one with wing fences.

Bf109G. With DB 601E engine.

Bf109G-0. With DB 601E engine.

Bf109G-1. Similar to G-0 with DB 605A-1 engine and GM-1 boost.

Bf109G-1/Trop. Tropical engine filters and one 20mm and two 15mm guns. Large fairings over breech blocks in engine cowling.

Bf109G-2. Pressure cabin deleted. Rearward firing MG 17s in a ventral position.

Bf109G-2/R1. Fighter bomber with provision for fuselage bomb and under-wing drop tanks.

Bf109G-3. Similar to G-1 but with different radio.

Bf109G-4. As G-3 but pressure cabin deleted.

Bf109G-5. More powerful DB 605D engine with larger supercharger plus MW-50 boost.

Bf109G-5/R2. Larger area fin and rudder of wooden construction.

Bf109G-5/U2. Wooden tailplane.

Bf109-G6. Powered by either DB605AM, AS, ASB, ASD or ASM. 30mm MK108 cannon over engine block, plus additional under wing MG 17s.

Bf109G-6/N. FuG 350 Naxos Z installed for homing in of H2S radar of British Pathfinder bombers. Armed with 15/20, 20mm cannon.

Bf109G-6/R1. Fighter bomber carrying a single 550lb SC250 or 1,100lb SC500 bomb on fuselage rack.

Bf109G-6/R2. Armed with 210mm WGr 21 Dodel rocket tube under each wing replacing cannons.Bf109G-6/R2 Trop. As R2 but with tropical equipment.

Bf109G-6/R4. Variation of 6/R2 with addition of twin 30mm cannon in under-wing gondolas.

Bf109G-6/R6. As above but 30mm cannon replaced by 20mm.

Bf109G-6/U2. Similar to 5/U2.

Bf109G-6/U4. Modified, semi-retractable tail wheel.

Bf109G-7. Definitive model of G6. Abandoned.

Bf109G-8. Reconnaissance variant with reduced armament.

Bf109G-10. DB 605 D engine with MW-50 boost. Maximum speed 428mph.

Bf109G-10/U4. Twin Mk.108 cannon in ventral tray. Or fixed overload fuel tank.

Bf109G-10/R1. Ventral carrier for single 551lb bomb or four 110lb SC50 bombs.

Bf109G-10/R2. 'Galland' hood, larger area fin and rudder and IFF radio.

Bf109G-10/R4. Twin MK 108 30mm cannon in under wing gondolas.

Bf109G-10/R6. As above but with 20mm cannon.

Bf109G-10/U4. Wooden empennage and increased area fin and rudder.

Bf109G-12. Two seat trainer version of G-1.

Bf109G-14. Fighter bomber. Could accept all DB 605 Series engines. One engine mounted 20mm MG 151/20 cannon; two 13mm MG 131 machine guns; plus additional twin 20mm MG 151/20s; or two WGr 21 rocket tubes; or single 250lb bomb.

Bf109G-14/U4. As above but with wooden empennage.

Bf109G-16. Additional armour, under fuselage bomb rack, and MG 151s in gondola.

Bf109H. High altitude fighter based on F Series.

Bf109H-0. pre-production H models.

Bf109H-1. DB601E engine.

Bf109H-2. Jumo 213 engine.

Bf109H-3. Jumo 213 engine.

Bf109H-5. With DB05L engine.

Bf109J. Spanish licence built project only.

Bf109K. Development of 109G-10.

Bf109K-0. DB 605D with GM-1 boost.

Bf109K. Pressure cabin. DB 605 ASCM/DCM and MW 50 boost. Armament of one 30mm and two 15mm cannon.

Bf109K-6. Armament of three 30mm and two 15mm cannon.

Bf109K-14. DB605L engine and MW 50 boost.

Bf109L. Project with Jumo 213E engine.

Bf109-S. Project with blown flaps.

Bf109T. Carrier fighter. Abandoned.

Bf109T-0. Ten conversions by Fiesler.

Bf109T. Conversion of 60 109K-1. Later to ground attack landplane.

Bf109T-2. As above.

Bf109TL. Project with two Jumo 109-0048 turbo-jets.

Bf109Z. Two 209F airframes joined by new centre section. Five cannon. Abandoned.

Bf109Z-1. Fighter bomber with two 30mm cannon and 2,200lb bomb load.

Bf Bf109Z-2. As above.

Bf109Z-3. Conversions of 109Z-1 with Jumo 213 engine.

Bf109Z-4. Conversion of 109Z-2 as above.

APPENDIX II

The Battle of Britain – a summary

The Battle of Britain – a summary of this crucial battle that was to determine air superiority for the remainder of the war. After Germany's successes in the Low Countries and France, Hitler was reluctant to invade England, whom he regarded as a stable force in the world. If he could persuade the politicians that he was content to rule Europe after having defeated Russia he would halt his activities. Through neutral sources he made his demands known but England's government rejected all approaches.

Not used to this rejection of what Hitler considered to be 'more than fair' demands, he decided, encouraged by Göring, to implement Operation *Sealion* – the invasion of Britain – and set a date in September 1940 as the last possible time for invasion. Had he done so in May/June that year there is a possibility that he could have achieved his ambition, but, by the time the first skirmishes took place with the harassing of the Channel convoys, it was too late.

Air Marshal Dowding, C-in-C Fighter Command, and his Group Commanders were confident they could face the mighty *Luftwaffe*. They had built up their squadron strengths, had radar and the Observer Corps to identify forthcoming attacks and, unbeknown to the Germans, Dowding also possessed information passed to him through the possession of a German 'Enigma' code machine. He knew in advance of when the raids would take place and the general strategy of Göring. But even with all these advantages the events of that summer forced England as close as it had ever been to defeat.

The battle that followed and lasted from July to October 1940 was without any doubt one of the most crucial of the war. Had England been defeated, and America stood alone without any allies, both Germany and Japan could have achieved world domination.

However, as stated above Hitler was in no mood to attempt such a move as he hoped that England too would sue for an armistice, and he even went as far to discuss with his generals how to run down the greatly enlarged Army and *Luftwaffe* until such time he could turn his ultimate intentions to his great aim, the destruction of his *bete noire* – Russia. He wanted to avoid the same situation that had faced his predecessor during the First World War, a two front situation.

He preferred to negotiate until such time he had gathered together sufficient, highly trained and equipped Army and Air Force for a swift campaign similar to that of Western Europe. In the meantime the *Luftwaffe* was to prevent any supplies and transport of goods reaching England through the English Channel.

Although the exact start of the battle was targeted for 13 August 1940, the *Luftwaffe* would take command of the Channel and seek to destroy all shipping, and generally harass RAF opposition if, and when, it appeared.

June passed quietly without any major operations taking place and Hitler still put out his peace feelers to Britain. Although Winston Churchill made it plain he would oppose any deal with Hitler there were members of his government and civil service who were quite prepared to talk with the various emissaries such as the Swedish and Swiss governments.

Overture, build-up to the battle

German troops occupied the Channel islands, as was to be expected being so near the French coast, on the last day of June, and 1 July witnessed a build-up of *Luftwaffe* attacks on Channel traffic. It was shortly after mid-day that the enemy made his first move, an attack on a Channel convoy. Radar identified a raid that was heading for Plymouth to the same position as the convoy and British fighters raced to the scene only to find that the attackers, a pack of Ju87 Stuka dive-bombers had completed their work and left the area. The first day's fighting ended with the RAF scoring a total of eleven *Luftwaffe* aircraft destroyed.

The following day was rather anti-climatic as rain and cloud spread over England and approximately 340 sorties were made by Fighter Command with no losses and 4 German aircraft destroyed.

The 3 July opened clear and sunny and enemy reconnaissance aircraft kept up a constant stream over the south coast. Later that day a small number of Dornier Do17s were seen to be heading for the Kentish airfields – Kenley, Edenbridge and West Malling. A single Spitfire was destroyed in an accident and 4 German aircraft, including a Bf109, were shot down.

Bf109E-4. Werk Nr 1253 *of II Gruppe, Jagdwschwader 54 'Grunherz'* during Battle of Britain. Note the elaborate camouflage scheme and also Dornier Do17 in background.

July 4. At breakfast time an early raid caught Fighter Command by surprise when a large group of Ju87s struck at Channel shipping and installations at Portland. They were gone before British fighters could reach the area. A second raid was launched against a convoy passing through the Straits of Dover at 2.00p.m. when a number of Dorniers escorted by Bf109s hit the convoy, again without any protection from British fighters. It appeared that Fighter Command was not reacting quickly enough.

After the raid the German fighters flew in over the coast looking for the opposing fighters and were met by Hurricanes. Much later in the day Hurricanes were despatched to face a group of Bf109s, it was a mistake as the enemy had the advantage of height but only one Hurricane was destroyed with the Bf109 group losing two.

The size of German attacks on the 5th was approximate to that on the 4th when rain obscured most the England and it was not until almost eight o'clock that radar directed Spitfires to a group of 109s and in the ensuing fighting one was shot down, making a total of four for the day. The Germans lost three aircraft.

Rain and low cloud in the South deterred the *Luftwaffe* from sending over any large groups of aircraft on the 6th, a Saturday. In spite of this there were two British aircraft casualties and five German, including Bf109s.

Sunday was to establish a pattern of raids that was pursued by the *Luftwaffe* and it appeared they had a psychological fixation about the English weekend. Fighter Command had supplied an escort of fighters for a Channel convoy and by mid-day Bf109s were harassing them. By late evening the main raid of the day was launched by a group of Dornier Do17s against the convoy which had neared Dover. Although British fighters were re-inforced over the shipping the bombers managed to sink one ship as the fighters fought off a group of Bf109s that had appeared on the scene. The loss tally for the day was seven British fighters and the same number of German bombers and fighters.

Monday 8 July saw a Channel convoy being shadowed by enemy reconnaissance aircraft as a prelude to a raid, and by 11.30a.m. a raid by Dornier Do17s was intercepted and turned away. Hurricanes flying top cover were bounced by 109s as they dived from altitude.

The German fighters operated in two separate formations, the *Schwarmen* of four fighters which left any group called the 'finger four' to then split into two, the *Rotten*. Four British fighters and seven German aircraft were destroyed during the day.

The pace of fighting increased on Tuesday 9 July and opened early in the day, but it wasn't until almost 3.30p.m. when a huge mass of German aircraft were met by Fighter Command fighters. The attacking fighters eventually split the large formation which dropped their bombs and turned away. A second raid prompted Park to move his squadrons to forward airfields.

This was not the final raid of the day as in early evening a raid by Ju87 and Bf109s attacked Portland once again and the British fighters sent to intercept clashed with the escorts, six of which were shot down.

An unusual number of convoys were present in the Channel on 10 July and the first was visited by a single Dornier Do17 on reconnaissance and supported by a heavy escort of 109s. British fighters attacked but could not reach the Dornier which got safely away.

The raid that took place a few hours later was anticipated when a number of squadrons were put on available and were scrambled at noon to meet a group of Dorniers escorted by Bf109s. There were over 100 aircraft taking part in the ensuing fight but a naval vessel was sunk. 4 German fighters were shot down for the loss of 1 British.

At 11a.m. on the morning of 11 July a group of Ju87s accompanied by 110s set course for Portland and when the raid was reported fighters were scrambled and headed for the area. The low level bombers were attacked, but jettisoned their bombs in the face of the aggressive British fighters. A second raid in the evening saw a group of German bombers heading for Portsmouth and they were met by fighters who quickly drove them off, shooting down eleven bombers and four fighters.

This tempo of raids continued throughout the following day when German forces attacked yet another convoy in early morning with Dorniers and Heinkels attacked and intercepted by Spitfires and Hurricanes. Fighter Command losses were eight with seven German. Sunday the 14th was quiet with just a single, small raid with few casualties on both sides.

A thick cloud base covered England on the following Monday with activity at a minimum and it was the same story for 15 July. Nothing much occurred until Thursday, the 18th, when the attacks started around nine o'clock with a raid on a convoy and there were similar clashes until dusk brought an end to daytime activity which saw the loss of five British and five German aircraft. The larger raids of later weeks were still to come, such as that of the following day, the 19th.

Early model Bf109E of JG26. This aircraft has a yellow nose, a green heart on the engine cowling and the 'Schlageter' emblem under the cockpit.

The Defiant fighter introduced during the Battle of Britian was different from the Spitfire and Hurricane in having no forward firing guns and a rear, four gun turret. On the 19th a squadron of Defiants was on patrol when they were bounced by Bf109s and in seconds the squadron's numbers were reduced to four. Portland again was a target for Ju87s but the day's events were black for the RAF with ten fighters shot down against a total of five Germans.

The pace of attacks quickened on the Saturday 20 July, and at dawn the first raid of the day approached the Thames Estuary consisting of Ju88s. At six in the evening a formation of Ju87s and Bf109s attacked a convoy and the dog fight swirled around the ships. The fighters broke formation when attacked. The score for the day favoured the RAF with its low losses of eight fighters as compared to the *Luftwaffe*'s sixteen.

On the following Sunday the skies were quiet with only minor clashes. This state of affairs continued until Wednesday the 24th. At eight in the morning the first convoy attack of the day took place when Dorniers attempted to bomb but were driven off by British fighters. Three hours later a large formation of Dorniers protected by 109s arrived to bomb the same convoy. Four Fighter Command aircraft were lost and eleven German.

The first action of Thursday, the 25th, took place at noon with a clash between British and German fighters near Dover. Forty-five minutes later a similar dog fight occurred, and as both sides broke off the engagement a large formation of Ju87s were spotted heading for a convoy accompanied by more Bf109s. The attack had been carefully planned to engage the British fighters first, and when they withdrew to refuel and re-arm, strike the convoy. Although a number of British fighters did manage to fly to the area the ships were bombed.

During the day the enemy concentrated on sending over groups of fighters, challenging the RAF to meet them on level terms, but the British controllers avoided that trap. Portland harbour had also been attacked several times during the day and No. 10 Group defended the base. Depite such activity only seven British fighters were destroyed as opposed to a total of German aircraft's total of seventeen.

Perhaps smarting from their wounds the enemy stayed away the following day except for reconnaissance flights – on the Friday and Saturday the skies were virtually empty. Sunday, the 28th, was busier, but not to a great extent. Park moved his squadrons to the forward airfields in anticipation of heavy action and at 1.30p.m. a raid was identified approaching Dover. However, the German bombers upon sighting the defending fighters, dropped their bombs and turned tail. Five German bombers and five Bf109s were shot down.

29 July dawned fine and clear and the fighter stations in southern England stood waiting for the events of the day to unfold. It started early, just after seven, with a convoy attacked by a large group of Ju87s and an even greater number of Bf109s. The attacked was vicious with no quarter given as aircraft after aircraft was destroyed. Despite the huge dog fights the RAF lost four fighters, the Germans thirteen bombers but nil fighters.

The month ended on the 31st with two days of little activity and the scene was set for a month in which Fighter Command would fight for its existence. *Adler Tag*, Eagle Day, was scheduled for 13 August with the launch of Operation *Sealion* scheduled for September. If England was not to be invaded then Fighter Command would have to hold, then beat the might of the *Luftwaffe*.

Crescendo

1 August 1940 started quietly but at 3p.m. a group of He111 bombers attacked Norwich, selecting the large aircraft factory there as its target. Another attack later the same day against a convoy off the Norfolk coast was driven off by Hurricanes. The RAF lost two, the Germans twenty. The following three days were fairly quiet with losses in Fighter Command of four fighters and the *Luftwaffe* eighteen aircraft.

On Monday 5 August the enemy concentrated his attacks against the Channel convoys and Tuesday saw numerous nuisance raids by fighters.

Thursday August 8 saw the opening of the second phase of the battle and has been regarded as the official first day. The major target for the day was a large, Channel convoy passing through the Straits of Dover and a major attack was launched against it at 10a.m. At noon a second blow was launched by a group of Ju87 dive-bombers escorted by fighters, and they were quickly identified by the Ventnor radar station.

The raiders were intercepted by British fighters as the convoy was bombed and soon the *Luftwaffe* fighters were tangling with the enemy. The raiders suffered many losses in that attack alone but there was more to come. It started at teatime when large formations of Stukas and Emils once again attacked the same convoy, which by now had suffered badly with many losses. The day was won by the *Luftwaffe* but at a heavy cost with thirteen bombers and thirteen fighters destroyed.

Friday the 9th was much quieter with scattered activity over the whole of the country as was Saturday the 10th which was the day on which *Adler Tag* was to have been launched.

Bf109E-7, II (Schlacht/LG 2), St Omer, France. Armed with four 110lb bombs.

On the following day, Sunday, Fighter Command was to feel for the first time the full weight of the *Luftwaffe*. It was a typical August summer day with fine weather prevailing and it opened when free chasing Bf109s swept in over Dover. But Park did not accept the bait despite the bombing of the town with little damage. This state of affairs continued for many hours and British fighters were scrambled in small groups.

At 9.45a.m. the pace of attacks increased when a formation of Ju88 bombers, Heinkels and Bf110s attacked the naval base at Portland. The German fighters arrived over the base first but did little except to circle overhead. However, British and German fighters swirled around and losses were high on both sides. This allowed the bombers to do their worst and bombs struck the base.

A Channel convoy was the next target for Dornier and Stuka bombers protected by Bf109s. However the raid was broken up and both sides withdrew with little to show for their efforts. Bad weather eventually brought an end to the day's activities. The huge total of thirty-two British fighters was a blow to both Dowding and Park who were aware that such losses could not be sustained for long, despite a loss of thirty-nine by the Germans. The Germans had greater reserves of men and machines.

Monday, the 12th, brought no respite to the tired young men of Fighter Command and the radar stations were selected as prime targets. The first to be attacked was that at Dover but the bombs missed the vital equipment despite damage to buildings. However, three stations went out of service for approximately six hours.

A number of airfields were next – Lympne and Hawkinge were bombed, with the former coming off the worst. A convoy passing Deal was attacked and this was followed by a massive raid which started just before lunch. Portsmouth was the main target and many military installations were hit, with part of the formation making a sharp turn to bomb Ventnor radar station which was reduced to a ruin.

Another airfield in Kent, Manston, was attacked during the afternoon by Dorniers which made a classic bombing sortie on the site. When the dust of battle had cleared it was discovered there was little serious damage. The day ended with a number of attacks by small groups of German aircraft, more of a nuisance than harmful. As was to be expected the tally for the day was high with twenty British fighters and thirty-one German aircraft down.

Adler Tag had been postponed to 13 August which started with attacks commencing at dawn with a formation of Dornier Do17 assembling to fly off and make for the Isle of Sheppey, over which they split into two groups to bomb Sheerness and Eastchurch airfield. Ju88s raided Odiham and Farnborough, but the real business of the day began after lunch.

Large formation of Ju88s, Bf110s and 109s had assembled and made for No. 10 Group territory to finally head for Southampton where they inflicted great damage to the port and surrounding areas. Another section of Ju88s headed once more for Portland while Stuka dive-bombers moved towards Middle Wallop to be met by a large number of British fighters.

Another section of this massive group of German aircraft gathered together and flew to Rochester intending to bomb the airfield, but were met by British fighters who scattered the invaders. However, a third group of Ju87s successfully broke through the defences and hit Detling airfield where they proceeded to demolish airfield buildings and a number of aircraft on the ground. It was a savage attack and the day's tally of losses was forty-one aircraft.

The pressure was sustained on the following day, Wednesday the 14th, but at a much lower level. It started with a group of Ju87s escorted by Bf109s reaching Dover and a section turned towards Manston to launch another attack during which many buildings were destroyed or damaged. The raids that followed in the afternoon were small and scattered and one target was Middle Wallop, the Army Co-operation field. Southampton was hit once again.

Among German losses of twenty-nine were many Ju87s and it must have been obvious to *Luftwaffe* chiefs that casualties among the force were mounting. Even when protected by fighters they were easy targets.

The date of 15 August was of significance to both Fighter Command and the British nation as the former had to fight one of the heaviest attacks during the fighting on the day. Just after 11a.m. large numbers of Ju87s were gathering together with their escorting Bf109s and the first target was Hawkinge but their luck was out as a large cloud of opposing fighters were waiting for them.

The Ju87s scattered and battle was joined between Spitfire and Hurricane and Bf109s. Lympne was struck again with many buildings demolished. A group of low flying Bf110s of the escort attacked Manston. After lunch Martlesham was attacked by Bf110s causing demolision of many buildings and some grounded aircraft. A large group of Dorniers and Bf109s approached Kent and attacked Rochester and Eastchurch airfields causing much damage to the Short factory at the former.

Hawkinge was attacked and the radar stations at Dover, Maidstone, Rye and Foreness bombed. The long suffering city of Southampton was bombed by Ju88s with a swarm of German aircraft attacking Middle Wallop and Odiham once again. Stukas struck at Portland under a cloud of Bf110s and 109s.

The last raids of the day were made on Biggin Hill and Kenley, by Bf110s carrying bombs and protect by Bf109s. Croydon also felt the weight of enemy bombs and many famous landmarks were destroyed. The final raid of the day saw West Malling attacked. The casualty rate for both sides was heavy – twenty-eight RAF fighters and forty German bombers and thirty-five fighters. The crisis was building for Park and Dowding.

Friday 16 August, was to be as bad as the previous day and the tired British pilots must have wondered how long they and their aircraft could go on. The day started at around 11a.m. when Dorniers bombed the new airfield at West Malling and was followed after lunch by strong formations with one striking at Hornchurch. They were successfully intercepted.

A more serious threat was the approach of a large group of enemy aircraft which attempted to bomb Harwell and Farnborough but it was thwarted upon meeting British fighters. At noon yet another group of Ju87 and Ju88s appeared and after splitting one group flew to Tangmere which was badly damaged and took some time to recover. Many grounded aeroplanes were destroyed and personnel killed.

Ventnor was once again a favourite target and so great was the damage caused that it went off the air for over a week. Lee on Solent was bombed and several aircraft destroyed,

Bf109E-4 as flown by Gruppen Adjutant, II./JG 54, Hermelinghen, 1940.

and Gosport followed. By late afternoon Fighter Command was wondering how long the day's attacks would go on. Hundreds of individual battles had been fought and despite their losses still the *Luftwaffe* despatched more aircraft.

The last raids of the day started when invading aircraft crossed the south coast and made for Biggin Hill but they were intercepted and turned away. Bf110s carrying bombs led the way with 109s providing cover.

It had started quietly enough as heavy cloud covered the country but by mid-day it was beginning to clear and the *Luftwaffe*'s intention was to smash the RAF and clear the way for invasion. An examination of the day's losses reveal how the battle swayed to and fro and resulted in a loss of twenty-eight for Fighter Command and thirty bombers plus eighteen fighters for the *Luftwaffe*'s attempt to swamp the defences.

Both sides rested on the Saturday but Sunday, the 18th, was a day of hard fighting which began when at mid-day the radar screens reported a huge build up of aircraft across the Channel. Nos 10 and 11 Groups brought all their aircraft to readiness while No. 12 went to available. Biggin Hill was attacked at one o'clock to be met by British fighters who shot several down. They were followed almost immediately by a group of Ju88s who flew in low but their bombs landed harmlessly.

Kenley was also under attack at the same time by a large force who destroyed buildings, cratered the airfield and destroyed a number of fighters on the ground. Casualties among ground staff was very high. A small group of the same bomber force turned for Croydon and then went on to West Malling, bombing both airfields.

After lunch, at two o'clock, a number of large formations of German aircraft appeared on the radar screens as they made their way across country to Poling radar station, Ford

Gosport and Thorney Island. While the Ju88s bombed Gosport unopposed the Ju87s approaching Poling were severely mauled by fighters.

In the evening another raid consisting of Bf109s struck at Manston, killing a number of ground staff and destroying fighters on the ground. It was obvious to the *Luftwaffe* Chiefs that the Ju87 could not operate successfully even with escorting fighters. They were too vulnerable. Records show that no less than seventeen were shot down in addition to twenty other bombers and thirty-one fighters, the majority of which were Bf109s.

The pressure slackened on the Monday 19 August, which was cloudy, with low casualties on both sides. Tuesday was similar and it was not until Wednesday that the raid strength increased with scattered raids taking place all over the country. But losses reported were low. Thursday and Friday was a period of almost calm conditions of weather and operations, but on Saturday the 24th the *Luftwaffe* returned with a vengeance.

Reconnaissance aircraft were followed by many attacks beginning at 8.30a.m. signalling the lull was over. Dorniers and Ju88, heavily escorted by Bf109s attacked many areas and did not cease until ninety minutes later. Hornchurch was hit as was Manston which was so badly damaged as to be rendered virtually inoperable except for minor duties.

A large force of bombers and fighters approached North Weald and the Dorniers, Ju88s and Heinkels smothered it with bombs. However, a number were turned away by determined attacks by British fighters. Aircraft from Leigh-Mallory's force arrived too late to contribute. A force of Ju88s bombed Ventor once again and then proceeded to make for Portsmouth causing heavy damage to the town and naval installations. The RAF lost twenty-five fighters and *Luftwaffe* lost a total of thirty-three of all types.

That weekend had started with a vengeance and was to continue throughout the following Sunday, the 25th. The weather was fine and clear but the *Luftwaffe* kept a low profile with sweeps of fighters up and down the Channel. The Fighter Command C-in-Cs refused to accept the bait knowing the futility of accepting battle on a fighter to fighter basis as a waste of crew and aircraft. At teatime the battle opened with a large raid of German aircraft approaching Weymouth Bay for the obvious target of Portland. Weymouth and Warmwell airfields were also attacked.

A small formation split from the major section and flew towards Southampton but was met by British fighters. At six in the evening another formation of German aircraft was forming across the Channel and attacked Dover. They were met by a horde of Spitfires and Hurricanes to which additional numbers were fed. The day ended with the *Luftwaffe* suffering the higher losses.

The event that was to turn the battle against the *Luftwaffe* occurred when Bomber Command retaliated to a German raid of the previous night by bombing Berlin. The order went out for the destruction of London by day and night, an impossible task for a force such as the *Luftwaffe* had.

From dawn of the 26th August the air war started once again with a raid by a bunch of German bombers, and their escorts fought a battle that raged across the Kent countryside. The German aircraft were successfully prevented from attacking Biggin Hill.

The fighting that followed began when raiders swung across Essex to reach and bomb Debden and escaped interception. However, a formation had the misfortune to run into many British fighters at the moment when their escorting Bf109s were making for their French bases. The bombers turned and followed, scattering their loads of bombs indiscriminately over Kentish towns.

Leigh-Mallory did despatch a number of his squadrons to aid No. 11 Group and managed to intercept a German formation. The next major raid was aimed at Portsmouth by Heinkels, Bf110s and 109s and these, too, were intercepted successfully. Losses for the day were high with twenty-seven British fighters going down and thirty-eight German aircraft lost.

Tuesday, 27th, was a day of little activity when compared with that of the previous day. Nevertheless there were casualties on both sides. The pressure again rose on the Wednesday, the 28th, and began at dawn with a raid by Dorniers with the intention of attacking Eastchurch and Rochford. Hurricanes barred their way and the Bf109s were soon heavily engaged leaving the bombers to reach Rochford. However, the airfield escaped major damage.

The German bombers were ordered to strike at Rochford once again and a group of Dorniers made the attempt only to be attacked by defending fighters. The next wave of aircraft consisted of German fighters and they fought it out with Spitfires and Hurricanes in an action that Fighter Command wanted to avoid. Park was alarmed at the cost, eighteen of his fighters against seventeen Germans.

Fighting was heavy once again the following day but started late in the day. At teatime, when enemy activity was observed across the Channel, Park brought his squadrons to readiness. A mass of bombers appeared protected by just as heavy a fighter screen of 500 Bf109s. The *Luftwaffe* was determined not only to protect its bombers but to soundly beat off and, if possible, destroy any fighter opposition.

Bf109E of JG26.

However, the British controllers recognised the tactics and refused to allow their fighters to engage the Bf109s. Nine British fighters were destroyed for an equal number of Germans. Although that amount of attrition was unacceptable Fighter Command had conducted itself well against such a mass of fighters that had trailed their cloak before them.

Friday 30 August saw the weekend approaching and the *Luftwaffe* was out in force. For Fighter Command it was the beginning of the final phase of the battle that would see one side or the other victorious. Soon after early light Dorniers and Bf110s appeared near Manston and were attacked by Hurricanes in what was a light attack, but the main battles were to come. Starting at around 10a.m. the first of many sweeps by Bf109s appeared. They were ignored and Fighter Command was now unwilling to match fighter for fighter.

One hour later the first heavy raid by bombers was launched with formations of bombers heavily escorted by 110 and 109s converging upon Kent. The first airfield to be attacked was Biggin Hill by Ju88s which made a successful strike.

From Kenley a number of Hurricanes were over Maidstone when they ran into a huge formation of 109s which were attacked with some success with no Hurricanes going down. A number of radar stations were off the air but were repaired within hours. At four in the afternoon a heavy raid consisting of formations of German aircraft attacked Luton, Slough and North Weald, Kenley and Biggin Hill airfields, with the latter heavily bombed by Ju88s once again. Twenty British fighters were destroyed and twenty-eight Germans of all types.

The *Luftwaffe* had saved everything up for the attacks of Saturday, the last day of August. The first raid was against North Weald and led by a great formation of Bf110s. There was also numerous Bf109s and both sides gave of their utmost. North Weald was bombed and Duxford attacked bringing No. 12 Group into action. Such was the weight of the defending fighters attack that the majority of the bombers dumped their loads and fled for the coast.

However, Debden was to feel the whole weight of an attack when a group of Dorniers slipped unobserved to the airfield and proceeded to bomb it. Despite the weight of bombs the airfield did escape serious damage. Another two airfields at Detling and Eastchurch were attacked with a number of Bf110s strafing the former.

All this action had taken place around breakfast time and more was to come with one large group of *Luftwaffe* aircraft racing again for Eastchurch. Biggin Hill and Croydon were the next victims when they were attacked around noon. As the bombers approached British fighters had them in their sights, but they, too, were attacked by escorting Bf109s.

Biggin Hill was severely hurt with most of its facilities put out of action. The Germans then turned their attention to Hornchurch where they bombed just after lunch. Both Biggin and Hornchruch were struck once again during attacks of early evening and the radar stations in Kent and Sussex damaged. The results of the day were damaging to both sides, except that the Germans had destroyed property in addition to aircraft. Thirty-four RAF fighters were destroyed, a damaging blow in spite of a total of forty German aircraft.

The Main Event – Fifteen Days of Destiny

The third week in September had been set for Operation *Sealion* when German forces including the Army and Navy would launch an all-out assault on the South Coast of England. The first strike would be by formations of bombers and fighters opening the invasion by attempting to prevent any opposition by the RAF and Royal Navy. The next stage would have been a mass attack by airborne forces dropping behind the front lines and wreaking havoc in the rear areas and securing strong points.

Luftwaffe Chiefs had their orders and 1 September opened at approximately ten o'clock with two groups of enemy bombers protected by formations of fighters hitting Biggin Hill, Detling, Eastchurch and London. This was a change of tactics as the attack on London would enable No. 12 Group the time and opportunity to assemble and unleash the mass squadron attack.

At one in the afternoon there followed a second strike against the airfields in Kent, but this raid did not approach London. The final attacks of the day were an attempt by the *Luftwaffe* to draw the British fighters into battle against the free ranging fighters. In this exercise the Bf110 was a lost cause as it needed the protection of the more agile Emil to enable it to perform.

Despite the severity of the day's fighting just thirteen British fighters were destroyed. The German forces lost eight. This was a situation Fighter Command could not sustain.

Monday 2 September. The first attacks of the day were launched after dawn to once again bomb four Kentish airfields, in addition a small group was to break away and bomb the town of Maidstone. At noon a much larger group of enemy aircraft were plotted approaching Dover and they were met by seventy-plus defending fighters.

Two further attacks concentrated on airfields and during this time a fierce battle took place, mainly between fighters. The enemy was succeeding in his strategy of forcing Fighter Command to commit more and more Spitfires and Hurricanes in one battle. The days returns of losses illustrates the complete picture of Fighter Command's agony. Twenty-five fighters destroyed, plus many trained pilots killed – which was the greater problem. Attackers twenty-four – still a plus for the Germans. However, this total did not include bomber losses.

Tuesday 3rd. The action started at around breakfast time and was confined to fighter movements which the RAF needed to avoid. However, when the fighters were identified in the main body of aircraft, a second, large, formation of bombers, plus escorting fighters, headed for North Weald. Park had to split his forces, directing the majority of his fighters to the aircraft attacking his airfield and attempting to ignore the large fighter mass. The day's losses showed a deficit by the Germans.

Wednesday 4th. This was a day of trial for Fighter Command as the *Luftwaffe* made it apparent they were intent on the destruction of the aircraft factories, without which the RAF would be defenceless. The day opened with attacks on airfields to prevent the defending fighters from operating to their full capacity. Another group bombed several towns in Kent including Reigate and Redhill, Canterbury and Faversham.

A further group made for the Hurricane facility at Brooklands but again mistook the Vickers factory as the main target. Civilian loss was horrific and the works were paralysed for several days before production could be resumed. The Short factory at Rochester was also a target. Losses were approximately even but this situation favoured the *Luftwaffe* who still had large reserves.

Thursday 5th. This was almost a repeat of the 4th except that the formations were smaller but there were more of them in an effort to split the defence. The oil storage facility at Thameshaven was bombed and fierce fires raged for hours providing a identification point for the German night bombers. It was, however, the first day for weeks that German losses were much higher than Fighter Command's.

Friday 6th. Another day which tested the defence to its limits as the Germans, apparently in an attempt to force the issue, almost swamped the defence system – but it held. It appeared as though the RAF was gaining the upper hand with German losses much higher. However, in the battle of attrition the *Luftwaffe* was still the stronger force.

Saturday 7th. On this historic day the *Luftwaffe* was seen to change its tactics. No longer would the airfields and aircraft factories be the main target. If the enemy considered he would bring the British people to the point of surrender he was to be disappointed. Their morale was steady and their defence system still adequate, as will be revealed.

The opening phases were a puzzle as the radar screens were blank but with plenty of warning the first movements appeared at three in the afternoon. It was a warm, languid day, almost peaceful, until the number of enemy squadrons approaching England were revealed. All three Fighter Groups were brought to readiness and the British fighter controllers waited until the *Luftwaffe* showed its hand. It was not long in coming as the attackers, more than 1,000 in total, flew steadily towards their intended target.

With all their aircraft in the air Nos 10 and 11 Group's fighters threw themselves against the towering mass of enemy fighters and bombers. However such was the weight of the attack that the *Luftwaffe* forced its way to London where it smothered the East End's dock system with high explosive.

No. 12 Group managed to get its squadrons into the air in time to frustrate Göring's plans to deal the death blow, little realising that the target was too large to bomb into submission. A second wave of bombers and fighters joined in the attack and Fighter Command faced a situation which it had endeavoured to avoid, total commitment of all its available fighters. Downing however still had slim reserves of American P-40 Hawks, Spitfires and Hurricanes which he was husbanding for the time when the land-sea invasion took place.

An examination of the loss and profit account tells its own story. Fighter Command thirty fighters, a heavy toll but not as bad as it could have been had it been confined to Nos 10 and 11 Groups. Enemy total forty. The attrition rate was still unequal.

Sunday the 8th was a quieter day with minimal losses on both sides and the opportunity to take stock. London was still the main target but the attacks slight.

Monday 9th. The pace of attacks increased on this day with aircraft factories and London as the main targets. In the afternoon a large formation of aircraft was attacked as it passed over Croydon with the result that this large suburb was subject to a hail of jettisoned bombs. It was a day when the *Luftwaffe* lost far more aircraft than the RAF.

Tuesday 10th. A quieter day with German bomber losses high when a group of RAF bombers struck at an airfield in France destroying a high number of German bombers. The RAF lost one. A good day for the defence.

Wednesday 11th. Action started after lunch when the weather cleared and the German bombers flew steadily towards London and the mass of British squadrons waiting for them. The Germans reached central London where there were few military targets, but the aim was to affect civilian morale. It was terror bombing in its basic state. The bombers were subject to the full weight of Fighter Command as their fighters had been left behind, and they suffered as a consequence. Thirty-one RAF fighters destroyed and twenty-eight German.

Thursday 12th. Little activity as, unknown to Dowding, the *Luftwaffe* was gearing itself up for the killing blow. However, this did not prevent a series of heavy raids on the Friday 13th against the airfields and radar stations in preparation for the weekend blow.

Saturday 14th. On this, and the following day both sides threw everything into the battle. London was the target initially with equal losses on each side as the *Luftwaffe* bombed it as the main target, as ordered by Göring.

Sunday 15th, the day of decision. It started quietly though Park and his cohorts knew that the fine weather was perfect for an attack. The entire force of Fighter Command stood waiting as the hours passed slowly by. Nerves were stretched and it was almost a relief when the control centres reported movement over France.

All fighter stations were called to readiness as the German force led by a mass of

A deceptive rural scene in France during the Battle of Britain. The Bf109E is camouflaged with tree decorations.

bombers headed for its target. No. 11 Group was the first in action as its fighters hurled themselves against the invaders, and as they neared the capital a second, huge formation of No. 12's fighters smashed into the flanks of the German force.

The German crews were devastated by the apparently unlimited number of RAF fighters and they jettisoned their bombs to make their way back to base with their escorts herding them along, trying protect their charges. The British fighters were quickly refuelled and re-armed and stood waiting for the renewed fight. It came within hours as the second raid formatted over France.

The German fighter force was now at strength, but the number of bombers to protect in the long flight from the coast to the capital stretched its resources to the limit. The first meeting of opposition forces took place over Kent with all of No. 11's fighters in the air. They harried the Germans all the way to London to break off with empty magazines.

However, No. 12 Group's fighters were ready and waiting. Over 20 full squadrons attacked the combined bomber and fighter force with such impact that it split apart. 300 plus British fighters wreaked their revenge as the fleeing Germans struggled first to jettisoned their bombs and then to join their fighter escorts as they all struggled towards their home bases. It was a mighty victory for Fighter Command.

The final raid of the day arrived just after seven in the evening when a force of fighters headed for Southampton to carry out an attack on the Supermarine factory that was building the Spitfire. What was odd was that this essential factory had attracted little attention before that day. The losses were counted as twenty-seven by the RAF and fifty-eight German. The odds were more even.

The following day saw enemy activity lessening as the *Luftwaffe* again sent over several raids consisting of fighters that were largely ignored. On Wednesday 18 September the enemy was content to commit fighters initially, but in the afternoon a group of Ju88 and Bf109s attacked the Chatham naval base and Rochester. Reaction by Park was swift and heavy as he threw his squadrons into the attack.

Thursday 19th. The RAF suffered no casualties while the Germans lost eight. Friday the 20th saw a similar situation with minimal losses for both sides.

Expected heavy weekend raids by the *Luftwaffe* did not materialise. Göring was avoiding committing his *Luftwaffe* to battering London as it was a lost cause. Also, Hitler had lost interest. The final day for launching Operation *Sealion* had passed and he had other interests for which he required the strength of the *Luftwaffe*, the invasion of his mortal foe, Russia.

The Battle of Britain was still a reality although the peak moments had passed. Tuesday 24 September started with large numbers of German fighters sweeping over southern England with losses of sixteen. The following day the RAF rose in some force in order to protect the aircraft factories, with Supermarine at Woolston being the main target. The factory took a number of bombs that halted production for several days until repairs were completed.

On the following day the Bristol factory at Filton was attacked, interrupting production of the Beaufighter, Blenheim and Beaufort, and, even more seriously, a blow against the engine factories. Plymouth and Portland also felt the weight of the *Luftwaffe*. A formation headed west to strike the Westland Yeovil factory, but on the way home they met a large formation of British fighters. Loss to Fighter Command was a low – five fighters as contrasted with German loss of fifteen.

Supermarine's Woolston facility was again attacked on Thursday 26th and this time damage was very severe leading to widespread dispersal of satellite factories producing major components and complete assembly at Eastleigh near Southampton and Castle Bromwich in the Midlands.

Friday 27th. Another day of heavy attacks that started at eight o'clock with a German fighter sweep intended to alert the RAF and commit its fighters. This was avoided and the fighter force was ready for the next raid that was aimed at London. The *Luftwaffe* had apparently ignored the experience of the battles on the 14th and 15th of the month and the force of Ju88s was attacked by a large number of British fighters who harried the bombers to the target and back with good results.

The Bristol factories were once again the target but the fighters were waiting and the invaders scattered. The pattern of attacks were now obvious, to bring the British fighters up to fight, exhaust them and then throw in the bombers against specialised targets. Fighter Command refused to play the *Luftwaffe* game and their defences were tightly controlled.

The ferocity of the day's fighting can be assessed with the number of casualties on both sides. RAF twenty-three, *Luftwaffe* fifty. The problem now was could the *Luftwaffe* continue to accept this situation. The squadrons in France were supposed to pound the British nation with no other object but to intimidate, while Hitler also wanted a strong *Luftwaffe* to continue the fight into Russia. The strain was to be too great.

The last weekend of September approached. Fighter Command braced itself and was ready when on the Saturday 28th the *Luftwaffe* gained the upper hand with the Germans losing four aircraft to the British fifteen. However, the supply of new men and aircraft was now sufficient to replace losses.

Sunday 29th was a day of action and the tables were turned yet again with greater *Luftwaffe* losses in comparison to those of the RAF.

Monday, the final day of September, saw a return to the mass attacks of the past few weeks. A large formation of bombers crossed the coast into Kent intending to reach London, but it was successfully intercepted and turned away. A sweep by *Luftwaffe* fighters was virtually ignored and Fighter Command stood ready for the next phase.

It came at mid-day when a strong force of bombers and fighters attempted to fight their way through to London but were met, and mauled, before they reached the outskirts. The final major raid of the day was another attack of Westlands, Yeovil. The RAF lost eighteen, *Luftwaffe* forty-two, a fair exchange.

The Autumn Anti-Climax

The shortening days of autumn brought about a change of tactics by the *Luftwaffe* and on 1 October the day opened with an attack on Portsmouth, but faced with a strong fighter force little damage was done by the bombers. The *Luftwaffe* made more use of the Bf109E-7 model which could carry a 250kg bomber on shackles under the fuselage. This led to the real fighter bomber, the Bf110, being dropped from many operations.

The same situation occurred on the following day, the 2nd, when fighters and fighter bombers dominated the scene. However, the score at the end of the day revealed that it was the *Luftwaffe* that was losing the war of attrition as more of their fighters were being shot down due to the larger squadrons of Fighter Command fighters. On this Wednesday the RAF lost three as opposed to seventeen Germans. The next day was quiet with little activity and it was only marginally busier on the Friday.

However, on the Saturday there was more activity and a number of attacks took place during the twelve-hour period. Three squadrons of Bf109s equipped with bombs headed for London, followed by a second squadron of Bf110s to attack the recently repaired West Malling airfield. A number of buildings were damaged but more damage was inflicted upon the German force. RAF losses were seven fighters and fifteen German, a ratio of two to one.

Sunday was extremely quiet but Monday 7th burst into life with numerous raids by German fighters over a wide area, including the Westland plant at Yeovil. A force of bombers led by escorting fighters bombed the factory with little damage but many personnel casualties. No. 12 Group fighters met the German force head-on and they lost twenty aircraft.

On the Tuesday 8th Park was forced to instruct his squadrons to fly at higher altitudes, in particular the Spitfire which now fought at 25,000ft, leaving the Hurricane to tackle any bomber at around 15,000ft. The Emil 109E-7 was the culprit even with its bomb load. On a day when Fighter Command lost only four fighters the German toll was much higher with thirteen missing.

The weight of raids continued throughout the period from Wednesday 9th until Saturday 12th when the RAF had a bad day with twelve fighters destroyed for five Germans.

On Tuesday 15 October the battle flared anew as the *Luftwaffe* contented itself with sending over fleets of Bf109 fighters to strike London with their single bombs. Park wanted to tackle them but was advised to husband his fighters should they be required later in the day. As the day closed to an early dusk the bombers appeared at seven in the evening and carried out a devastating raid as the day fighter force of Fighter Command could not operate at squadron level in the dark. Despite this losses were high on both sides.

The 16th was a better day for the RAF with just two fighters destroyed for a toll of fourteen of the enemy. On the 17th the same weight of attacks continued, most by the Emil, and Friday, the 18th, witnessed the slaughter of Ju88s.

Saturday and Sunday were quiet with small numbers of nuisance raids, that situation prevailing until Friday 25th when the airfield of Montrose was hit sustaining damage and personnel casualties. Sixteen RAF fighters were shot down against a loss of twenty-one German, most of these being Bf109E-7s.

Saturday 28th was a heavy day of fighting and on the following day it flared again with numerous hit and run tactics by German fighters which caused alarm among the coastal towns though little damage. Monday 28th was a good day for the RAF who suffered a single casualty in return for the loss of thirteen *Luftwaffe* aircraft.

The final three days of October were days of much activity, as though the *Luftwaffe* wanted to end its summer/autumn offensive with a flourish. Over the three days the RAF lost eighteen aircraft and the *Luftwaffe* thirty-seven, much to their chagrin.

With the daytime battle over, the Germans continued to batter the cities of England until the Russian front drew the *Luftwaffe* away. Although no truly accurate figures of RAF fighters shot down by Emils during the Battle of Britain exists, such records that do exist enable certain conclusions to be drawn and suggest that about 70% of Hurricanes, 60% of Spitfires and 95% of Defiants lost were due to the guns of the Emil.

For a total loss in the air of approximately 587 Emils during the four months of the battle, the German single-seat fighters destroyed approximately 380 Hurricanes, 210 Spitfires, 18 Defiants and 27 Blenheim fighters – a grand total of 635. The remaining 300 aircraft fell to the guns other German types. Göring and Hitler had taken the fatal decision to concentrate daylight attacks on London from 7 September – at a critical moment in the fight for England. That was the measure of the quality of the Bf109E as a weapon in that critical battle.